Boost Your Brain

The New Art and Science Behind Enhanced Brain Performance

Majid Fotuhi, M.D., Ph.D.

and

Christina Breda Antoniades

HarperOne
An Imprint of HarperCollins *Publishers*

HarperOne

This book is written as a source of information only. The information contained in this book should by no means be considered a substitute for the advice of a qualified medical professional, who should always be consulted before beginning any new health program.

Products, pictures, trademarks, and trademark names are used throughout this book to describe and inform the reader about various proprietary products that are owned by third parties. No endorsement of the information contained in this book is given by the owners of such products and trademarks and no endorsement is implied by the inclusion of products, pictures, or trademarks in this book.

All illustrations are used courtesy of the author.

HarperCollins website: http://www.harpercollins.com
HarperCollins®, ®, and HarperOne™ are trademarks of HarperCollins Publishers.

FIRST HARPERCOLLINS PAPERBACK EDITION PUBLISHED IN 2014

Designed by Laurel Muller

Library of Congress Cataloging-in-Publication Data
Fotuhi, Majid.
Boost your brain : the new art and science behind enhanced brain performance / Majid Fotuhi, M.D. and Christina Breda Antoniades. – First edition.
pages cm
ISBN 978-0-06-219929-4
1. Mental illness–Prevention–Popular works. 2. Mental health–Popular works. 3. Brain–Popular works. I. Antoniades, Christina Breda. II. Title.
RA790.F65 2013
612.8'2–dc23 2013005876

14 15 16 17 18 RRD(H) 10 9 8 7 6 5 4 3 2 1

To our children,

Nora and Maya

and

Vasili, Marguerite, and Eleni,

whose amazing brains are only at the start of a wondrous journey

Contents

Foreword

This book may be the most important you'll ever read. Certainly what Dr. Fotuhi teaches in this book has changed my life. Surprised? I was.

What Dr. Fotuhi teaches has given me new enthusiasm for life: I now feel confident about my memory and brain performance. Learning how to memorize a list of twenty things, and then forty—forward and backward—in just about an hour, made me realize that "I could do it," something I doubted strongly just a few weeks ago. I soon applied my newly discovered strategy to remember names at a dinner party and was thrilled to discover that I am now actually good at remembering names, something I always felt was my weakness. Learning about the cutting-edge research on how simple lifestyle changes can increase brain size provided new incentive to be more conscious about everyday choices. The information in this book supercharged my energy and attitude toward brain aging.

If you've read the *RealAge* or *YOU* books or caught me talking about aging on TV, you know I strongly believe that when it comes to longevity and living well, your "RealAge" depends less on what the calendar tells you and more on how you choose to live your life. Science backs this up: there's clear evidence that certain choices lead you to better health, which can make you years—even decades—younger than your chronological age.

I've put the science to work for myself: my RealAge puts me at forty-six, far younger than my calendar age of sixty-seven. But before reading *Boost Your Brain,* I'll admit I was beginning to worry I'd hit the tipping point for brain function. Yes, I knew tricks—like taking a daily dose of algal DHA—to keep the "house" I live in much younger than its chronological age, but I was afraid my electrical system—my brain—was constructed of fraying copper wires that couldn't be upgraded to serve up the power the modern computer-world demands. Dr. Fotuhi changed my attitude and understanding. He taught me new secrets, like how to do the exercises I was already doing but in a way to increase my brain size and memory. And Dr. Fotuhi convinced me to reinstitute some of the steps I'd started to skip.

How much can these steps really matter? A lot. As a matter of fact, how long you live once you pass the age of thirty (genes play a large role early in life) and how healthy you are while alive is largely determined by the choices you make. By the time you reach the age of fifty, lifestyle factors account for a whopping 70 percent of your health and longevity, with genetics accounting for just 30 percent.

This may well fly in the face of the way you've always viewed aging. Many people—even doctors!—think that you reach your peak quality of life in your mid- to late twenties and then inevitably begin a long slow decline that lasts the rest of your life. There's a reason that perception has held sway for so long: it's true. Or at least, it's true for most. Provided you do nothing about it, you can expect every function in your body—your lung function, heart function, muscle mass, bone mass, even your IQ—to decline by 5 percent every decade, beginning right around your thirtieth birthday.

In public speeches I sometimes illustrate this with a rather telling slide that shows the results of a study of Harvard physicians. The study, which launched in 1952, tracked the physicians' IQs over their life span and the resulting downward curve shows that on average their brain function fell by 5 percent every ten years. If you're over the age of thirty you might find this discouraging. But you'll probably admit that it's also not that surprising. After all, don't most people feel more forgetful and mentally fuzzy as the years pile on?

If you drill into the data from that study of Harvard physicians, though, you'll find the curve also offers some promising news. Looking beyond the average you'll find a group of people who seem to defy nature: 25 percent of the physicians don't show any decrease in their brain function as they age. That's right: they're as sharp at eighty as they were at forty.

The question on everyone's mind when they see that slide is, understandably, "How do I join that club?"

You'll be happy to hear that although it's a small club, it's not really that exclusive. You don't need to have won the genetic lottery to get in. You don't need to be a Harvard physician, or wealthy, or well connected. Admittance largely comes down to the lifestyle and choices you make in your life.

At the Cleveland Clinic, we've ushered thousands of people into this club by developing programs that help employees improve their fitness, eat a healthier diet, manage their stress, and quit using tobacco. The programs have been wildly successful, helping our employees become many years—or even decades—younger than they were.

I've seen firsthand the incredible impact moderate lifestyle changes can make, even when they're started later in life. It's part of the reason I was so intrigued when I first heard of a Baltimore-based doctor with a novel approach to helping improve patients' memory and thinking.

At his Brain Center, Dr. Fotuhi has pioneered a whole-body approach to brain health that incorporates cutting-edge, neuroscience-based strategies specifically designed to maximize mental function. Through his twelve-week brain fitness program—a sort of boot camp for the brain—Dr. Fotuhi helps his patients grow (yes, that is the right term) their brains and improve their memory and thinking. (As a bonus, they almost always improve their overall health as well.)

In *Boost Your Brain,* Dr. Fotuhi brings that twelve-week program to readers: analyzing, distilling, and translating into plain English the groundbreaking discoveries that allow you to make your brain years or decades younger and really tap in to its amazing potential. Beginning with an understandable introduction to the basics of brain science, Dr. Fotuhi elegantly explains an idea that even some doctors have yet to fully grasp: the adult brain is incredibly malleable, growing (and shrinking) based on your behaviors and choices. That's incredibly important given, as Dr. Fotuhi will tell you, that the brain is the only organ in the body where size matters. Taking steps to help the brain grow—whether you're thirty, sixty, or ninety—not only helps your brain function better, but keeps it younger longer.

Boost Your Brain details the science behind four key brain boosters and guides you to custom-create your own twelve-week brain fitness program. Along with his step-by-step prescription for brain growth, Dr. Fotuhi delivers more real-world advice on how to avoid life's many brain shrinkers—from excessive stress, to hypertension, to chronic insomnia, and more.

His program and this book come at a time when chronic health conditions like diabetes, unmanaged stress, and obesity—the very problems that keep us out of the 25-percent club—are poised to reach unprecedented proportions. And it arrives on the leading edge of a wave of scientific discovery that allows experts like Dr. Fotuhi to understand the workings of the brain like never before.

If you've ever thought you'd love to get back the brain you once had, or have a forty-year-old brain when you're sixty, this is your chance. You're not bound to the brain your birthday says you should have. With *Boost Your Brain,* a bigger, younger brain is yours for the taking. Grab hold. I have, and you can, too!

Michael F. Roizen, M.D.

PART I

The Power Within You

Introduction

Your Bigger, Better Brain Is Within Reach

You've just walked into a meeting and you're confronted with a familiar face. You've met him at least a half dozen times, chatted with him about the weather, the annual summer picnic, and your company's latest sales figures. You're about to say hello when, with a sinking feeling, you realize you have no idea what his name is. Or rather, you know it. You know you know it. But as you walk toward him with a rising sense of panic, your brain stubbornly refuses to deliver the goods. Jeff? Jim? Jordan? You think to yourself, not for the first time, *Why can't I just remember?*

If it were just this flub with Jeff or Jim or Jordan (whose real name, you later discover, is Barry), you wouldn't be concerned. But it's not. For longer than you can remember (no pun intended) you've been feeling forgetful, a little slow on the uptake, slightly short on creativity. It's annoying—even frightening. You joke about your shredded memory or blame the latest lapse on the all-nighter you pulled last week, but sometimes you wonder, *Will I always struggle like this?*

If it offers any comfort, you should know that you are by no means alone in your worries. Most people feel their brains aren't functioning as well as they used to, or as well as they should. And they're right. In part, that's thanks to a basic biological reality: the human brain shrinks with age. Over time, brain cells known as neurons shrink or die. The contacts—or synapses—between neurons are lost, the communication highways that crisscross the brain deteriorate, and blood vessels wither. This happens first at the microscopic level and then, eventually, to such a degree that it can be seen on magnetic resonance imaging (MRI) with the naked eye. Such brain shrinkage is so commonplace that I, and most neuroradiologists, can ballpark a person's age simply by looking at his or her MRI.

As you can probably guess, such shrinkage comes with a cost, as numerous experts, including myself, have outlined in a host of scientific journals:

a smaller brain means poorer cognitive performance.[1] Shrinkage in the front of the brain makes us less focused and slower in solving complicated puzzles, making decisions, or planning for the future. (Later in life, it even slows the pace of our talking, walking, and making simple calculations.) Shrinkage in the memory areas of the brain makes it more difficult for us to recall names, phone numbers, or directions.

What's not so obvious is that brain shrinkage isn't just a worry for late life. The earliest footprints of atrophy in the brain begin to accrue as early as our forties, around the same time many people begin to feel they're not as mentally sharp as they once were.

There's an even more striking reality: no matter what their age, the vast majority of people aren't functioning at their full potential. Not even close. They may be operating on six cylinders in their midtwenties, or four cylinders in their seventies, but what they don't realize is that, at any age, they could be functioning at a markedly higher level—powering through life on eight cylinders. They place too much blame for their lapses on their hectic lives or their age and rarely even attempt to fully tap into their brains' power. Anybody can learn a language in their fifties if they have to, for example. But most decline to even try. "I'm too old to learn that," they'll complain. But the fact is, they *could* learn a new language, if they had enough of an incentive. Imagine what would happen if they were offered a $10 million prize for learning to speak Italian. I assure you most would be speaking fluently in mere months.

This isn't just a matter of willpower either. The amazing truth—one that the field of neuroscience has begun to clearly understand only in recent years—is that parts of the human brain, especially those parts important for memory, attention, and problem solving, have innate plasticity: the ability to change. Just as they can shrink, they can also *grow,* getting thicker, denser, and larger. The result is what I call "enhanced brain performance," a brain that functions at its highest level.

The process of growing these parts of the brain is akin to upgrading a battered six-cylinder engine to a gleaming eight-cylinder. And here's the kicker: parts of the additional cylinders we need are delivered to all of us every day but are ignored. Not just by some people but by most. They know that eating certain foods or exercising can help their six-cylinder engines run more smoothly, but they have no idea that with a concerted effort they could actually grow their brains. And they don't know that those two extra cylinders hold the key to reclaiming clarity and creativity and boosting memory—for life.

If that's news to you too, don't feel bad. This is cutting-edge science, new to many even in the field of neurology. My own understanding of this incredible reality has come after thirty years of studying and publishing scientific research, teaching at Johns Hopkins University and Harvard Medical School, and putting into practice the latest discoveries in neuroscience.

It was, in fact, my scholarly research into the topic that inspired me to write this book. As I pored over hundreds of scientific studies, I was struck by an undeniable truth: not only can the brain grow, at any age, it can do so within mere weeks or months, rather than years or decades.[2] Excited about what I'd learned, and eager to share it, I began to put together the outline of this book.

But How?

As it happens, the brain's incredible plasticity is most dramatic in the hippocampus, a thumb-size structure important for short-term memory, where we now know that new neurons are born every day, even in adulthood, through a process called neurogenesis. But plasticity is also evident across the brain, where growth takes place in the form of newly born synapses, larger and more intricate communication networks, and a vibrant system of newly formed blood vessels (born in a process called angiogenesis).

You'll learn soon enough why and how this happens, but the short story is that such growth in the number of new brain cells, synapses, connections, and blood vessels depends on three critical elements: increased oxygen flow to the brain, increased levels of a critical protein called "brain-derived neurotrophic factor" (BDNF), and the harmonious regulation of brain wave activity. Alone, and especially in concert, these three elements fuel growth in the brain.

The results can be dramatic—just like shrinkage, brain growth can be so significant that it can be seen on MRI with the human eye. By adding synapses, bolstering the brain's highways, breathing new life into blood vessels, and even growing new neurons, we literally reshape the brain and build a bigger "brain reserve," which enhances brain performance—at any age.

If ever there were a fountain of youth, this is it. In fact, I believe that most people can stave off brain aging, making their brains younger—by as much as five or ten years—and boosting their cognitive performance. It's the difference between polishing those four cylinders until they purr (as past approaches to brain health have relied on) and powering up, day in and day out, on all eight.

Bigger *Is* Better

Everything you do is a reflection of the brain you have. Whether you're troubleshooting a technical problem at work, weighing a new job offer, or merely wording a message on a friend's Facebook wall, how—and how well—you function cognitively is a direct reflection of your brain's size and health.

So, whether your brain is shrinking or growing—be the changes microscopic or macroscopic—*size matters*. And when it comes to peak brain performance, bigger is undeniably better.

In the hippocampus, growth brings with it improvements in short-term memory. Similarly, larger frontal lobes are tied to improved decision making and processing speed. A larger network of connectivity in your brain's intricate communication system, meanwhile, improves creativity and the ability to solve abstract puzzles. In short, a bigger brain enhances brain performance in three key areas:

Memory

Research has shown that your ability to learn and remember is tied to a bigger hippocampus. By taking simple steps to create synapses, promote neurogenesis and angiogenesis, and expand your brain's highways, you can increase the size of your hippocampus and sharpen your memory in as little as three months. One example is practicing memory exercises, an endeavor I think of as flexing the "memory muscle" in your hippocampus. Such exercises will improve your memory in the short term and make you a master of memorization as your hippocampus grows.

Clarity

We all have moments of mental fuzziness. Faced with a problem at work, you might fail to connect the dots that lead to a rapid solution. In a social situation, you might miss a cue or make a gaffe. Later, when the best answer strikes you (too late!), you berate yourself: *Why didn't I think of that then?* Even at its best, your brain can't banish these moments entirely. But a toned-up, bigger brain will operate more efficiently and quickly, giving you clarity of thought and a better ability to connect those dots on the fly.

Creativity

I'm not promising that you'll suddenly morph into Matisse or write a bestseller tomorrow. (Although, who knows?) But I can guarantee that growing your brain will translate into a greater ability to solve problems and, therefore, enhance your creativity. If you're an artist, that might mean solving the problem of precisely how to blend hues to achieve a masterful effect. For a doctor, creativity might mean puzzling through a complicated medical case to come up with the correct diagnosis. For a parent, it may come down to ably juggling the schedules of a family of five—and still finding time to make it to a book club or a date night with the spouse. We all problem-solve hundreds of times a day in every area of our lives, and our ability to think creatively goes a long way toward making us more successful at home, at work, and at play.

Your bigger brain can deliver improvements in these three areas—in just twelve weeks. But you should know, too, that the benefits of a bigger brain last a lifetime. In fact, growth in the hippocampus, in particular, has been linked to a reduced risk of developing the symptoms of Alzheimer's disease late in life. In one study, people with larger hippocampi were less likely to show signs of dementia, even when their brains harbored a significant load of Alzheimer's pathology.[3]

The Brain Center in a Book

When patients come to see me at the NeurExpand Brain Center, they often report being forgetful, foggy, and slow. "I'm not as sharp as I used to be," they tell me. Some are young. Some are old. They know they're not performing at their peak, but they have no idea why, or how to get their mental mojo back. Most—if not all—worry that they're experiencing the earliest signs of Alzheimer's disease.

I'm often able to tell them that their cognitive issues aren't Alzheimer's disease but are instead related to changes in the size of their brains. Typically, they are amazed. They've known that they should eat better, lose weight, and watch their blood pressure and cholesterol. They know that stroke and heart disease are devastating conditions—and often avoidable—and that stress might be burdening their bodies in ways they don't even understand. "But no one ever told me that these things could be shrinking my brain," they say. And nobody, without a doubt, has ever told them how to *grow* their brains.

Suddenly, in place of an abstract notion of improving their health, they have a very real goal. It's as if a lightbulb snaps on. Those bacon double cheeseburgers they load up on aren't just making them unhealthy; they're shrinking their brains. And the once-treadmill-now-clothing-rack in the corner of their basements? That, they realize for the first time, represents a missed opportunity not just to minimize their waistlines, but also to boost their brains' size by generating new neurons and rejuvenating existing ones. When I describe for them the science behind meditation or brain wave training, they're able to visualize trading the choppy waves of their stressed-out brains for the smoother, healthier waves of a brain humming along in the optimal frequency range.

They get it. And they're almost always eager to get started. They want to hit the gym, revamp their diets, sleep better, dial down the stress in their lives, and retrain their brains. But what exactly do they need to do? How often? How much? And for how long? To make it work, they need details.

Most likely no one has ever told them how to care for their brains, so at the Brain Center I start by educating them on the basics of brain health—giving them a condensed version of the book you're about to read. They learn about all the factors that contribute to brain size and fitness, and about the vital role played by oxygen, BDNF, and healthy brain activity. I explain that the one-size-fits-all approach will not work and that we need to assess each person's current brain health and make a plan with that in mind.

After considering their current brain health (you'll calculate your brain fitness level in chapter 3), I develop for them personalized treatment plans, which incorporate the expertise of my seasoned team of health care professionals. In addition to being seen by me, they spend time with a neuropsychologist, a dietitian, an exercise physiologist, a sleep specialist, a brain wave trainer, and a "brain coach." They may also attend memory boot camp sessions—learning tips to stretch their memorization skills—or drumming sessions aimed at training their brain waves to oscillate in harmony. They might even enroll in a neuro-friendly physical fitness program I've developed.

All of it happens within the walls of my Brain Center and all with one goal in mind: enhancing the brain's performance by expanding its size.

As my patients can tell you, it works. Every day I hear from patients who report significant improvement in memory, clarity, and creativity. I hope to one day bring the services provided by my Brain Center to people of all ages nationally. It's a concept I believe will revolutionize the way we view brain health and fitness as well as how we prevent and treat the memory and cognitive problems that typically come with aging.

That's the vision. And while I can't magically squeeze a full-service Brain Center into the pages of this book, you will find within its pages the tools you need to grow your brain—with benefits that begin now and carry on into the future.

In chapter 3, you'll begin the process of creating your own twelve-week brain performance enhancement plan. By chapter 8, you'll know enough about the science behind a bigger brain to start putting your knowledge into action. Over a three-month period, you'll implement my brain fitness strategies and track your progress, rating yourself each week in critical brain fitness categories. Within weeks, you will have vastly enhanced your brain performance and will feel the effects in your daily life. In fact, you'll likely feel better than you have in years.

But I want more than that for you. I want this book to change your life.

That's a tall order. It will take change and resolve on your part—a mindset you'll be far more likely to embrace once you truly understand what your actions (or inactions) do for your brain.

By the time you turn the final page of this book and complete your twelve-week brain fitness program, you *will* have a bigger brain. I hope, too, that you'll have radically altered the way you think about what you eat, how you spend your day (and night), and the choices you make that affect your emotional well-being.

It all goes back to that near-universal worry of yours. Will you always struggle like this? You don't have to. The power to change is within you. If you embrace the opportunity today and every day, you *will* grow your brain. It's as simple as that.

CHAPTER ONE

Your Marvelous Mind: A Growing Machine

MY FASCINATION WITH the brain—how it develops in utero and then in childhood, how it matures in early adulthood and then begins to shrink by midlife, how it shrivels in late life—took root during my days as a student at Harvard Medical School. I have cherished memories of those days. Some, though, stand out vividly in my mind—like the day I first delivered a baby.

It could have been just another day in the life of a medical student. Certainly, it started much like any other, with me waking before dawn in my small Coolidge Corner apartment, bundling up against the frigid cold, and then setting out for my two-mile bicycle commute through the snowy streets of Boston. My destination was the maternity ward at Beth Israel Medical Center, where I was midway through a three-month obstetrics rotation as part of my basic clinical training.

As I had every day for the prior six weeks, I would spend the next twelve hours attending to patients in various stages of labor, checking their progress, and answering their questions. A third-year medical student, I had already acquired a reasonable base of knowledge about the human body, thanks to long days of lectures followed by late nights cramming my brain with the details of everything from the biochemical processes of sugar metabolism, to cancer-causing cell mutations, to the bioethics of medicine. In my second year I had completed a general surgery rotation, starting first as an observer and then slowly, slowly carrying out minor surgical tasks of increasing difficulty, learning by doing under close supervision. In my obstetrics rotation, I'd already observed as many as two hundred deliveries and assisted in twenty.

I had started medical school with a deeper well of medical knowledge than the average student—at least when it came to brain-related matters—having taken the unusual route of completing my Ph.D. in neuroscience at Johns Hopkins University before I started medical school. I was actually in the

M.D. program at Harvard by way of a teaching scholarship offered through Harvard-MIT Division of Health Sciences and Technology, which put me in the unlikely position of teaching neuroanatomy to my medical school classmates while I was a student myself. I knew neuroscience. But for medical conditions below the neck, I was in much the same position as my classmates: nervous, but also eager to learn everything I could about the human body.

So, it was with equal parts uncertainty and excitement that I greeted my obstetrics professor as she sauntered over to me in the delivery room that day. Motioning to a nineteen-year-old woman who was nearing the final stages of labor, she said, "You'll deliver this one," turning an ordinary day into one I will never forget.

In the delivery room, I was gowned and gloved and, I'll admit, sweating profusely as the baby began her trip down the birth canal. A new life was entering the world and I didn't want to mess it up. Thankfully, everything went according to plan, and after the mother-to-be gave a few final pushes, I eased the baby into my hands. Now, this is an incredible moment even after you've experienced it a few times, but I'll never forget the feeling I had sitting there, amid the happy tears of the new mom and her family (and even a few of my own), holding that baby girl, who would soon be named Sara.

Nine months ago she was a microscopic bundle of rapidly dividing cells—first two, then four, then eight, sixteen, and on (into the trillions!). How could she have grown into this bawling, squirming, perfectly formed human? Along the way, a million things could have gone wrong. And yet, here she was. With eyes that one day would be able to distinguish several thousand shades of color, a complex auditory system capable of transforming vibration in the air into sound (and turning those sounds into intelligible thoughts and concepts), and a heart that would beat some hundred thousand times a day, every day.

One year later, having spent time training in the pediatrics, psychiatry, cardiology, and medicine wards, I was doing another rotation in my life as a medical student, this time in neuropathology, an elective whose choice reflected my fascination with the human brain. Neuropathology, the study of diseases of the central nervous system, involves performing autopsies to detect disease in the brain (both by examining it first with the naked eye and later under the microscope), and I was as eager to jump in as I had been on my obstetrics rotation. I was grateful, too, to finally spend weeks focused on exploring the brains of patients who had Parkinson's disease, multiple sclerosis, head trauma, and Alzheimer's disease. By comparing their brains to those of otherwise healthy elderly people who had died of natural causes, I knew I would gain a new understanding of the workings of the brain and how it changes with aging and disease.

Just as I had for the OB rotation, I progressed in steps, first watching someone else perform a brain autopsy, then inspecting an already-prepared brain, and finally, handling the delicate procedure of removing the brain from the skull. But unlike in obstetrics, neuropathology takes place in a morgue, a cold, sterile room—think bare walls, metal tables, and concrete floors—inevitably tucked away in the bowels of the hospital building. There's another big difference: in neuropathology, patients don't cry or squirm or come to harm if you make a wrong move.

Still, the brain can't be properly studied if it's handled roughly or, God forbid, dropped. And you only get one shot at successfully removing it from the skull.

My moment to perform a brain autopsy finally came one weekend morning with a call from the morgue. "If you want to do this one, come on in now," the professor I was working with said. I hopped on my bicycle, pedaled to another Harvard teaching hospital, Massachusetts General Hospital, and quickly changed into hospital scrubs. Awaiting me in the basement morgue was the body of an eighty-five-year-old woman who had died of aspiration pneumonia and had suffered from dementia. Her name was Mrs. Grey.

By this point, I had seen enough cadavers and exposed brains that I was quite comfortable entering the morgue and preparing for the autopsy. Still, as I donned my protective gown, face mask, and hood, I felt a nervous anticipation that wasn't too unlike those days in the delivery room.

To remove the brain from the skull, the coroner and I first had to cut away the top of the skull, just above the eyebrow, with the help of an electric saw. There is some delicacy needed: slice too deeply and I'd destroy the delicate grey matter within. Once the skull was cut, I set the bowl-like top piece aside and gently pulled the brain outward, making space to slice away its attachment to the spinal cord. I then eased Mrs. Grey's brain out and let it fall gently into my cupped hands.

A brain has the consistency of Jell-O and, like a baby, it is fragile—and a little slippery. In a moment, I would weigh it, examine it visually, and then place it in a formalin-filled bucket to preserve and harden it so that it could be dissected and observed under a microscope. Even without a microscope, though, I could see the ravages of time—instead of the plump peaks and minimal valleys of a young, lush brain, Mrs. Grey's had deep ridges separated by wide spaces.

As I cradled Mrs. Grey's brain in my hands, I remember marveling in much the same way I had on the day Sara was born. In my hands were three pounds of cells that had powered an entire life. It was the very essence of that life: the neurons and connections that spurred Mrs. Grey to volunteer at the library on weekends, smile when she saw her grandson's face, or get grumpy when she was stuck in traffic.

In between the moment she'd first seen those bright delivery-room lights and the moment her last neuron had fired, every experience her brain had encountered shaped the person she became. And yet, no one at the time would have suggested that how she lived her life as an adult had substantially altered the size and structure of her brain. At the time, in the mid-1990s, the human brain was widely believed to finish its development and become fixed in structure by childhood. After that, there would be only one inevitable possibility for structural change: the brain would shrink with age. That's what everyone thought. But, as you'll soon learn, "everyone" was undeniably wrong.

Your Brain, Baby

What happened to Mrs. Grey's brain in those eighty-five years between birth and death? And how did changes inside her brain affect her cognitive abilities throughout life? How might she have steered those changes in one direction or another–and to what end?

Mrs. Grey's story started the same way that Sara's had: with a tiny clump of cells. I'll stick with the simple version of brain development and aging, but even simplified it's an incredible tale. Human life, after all, begins as one cell and ends with some hundred billion in the brain alone.

We can start the story of Sara's brain by taking a peek inside her mom's uterus at about the third week of gestation. Remember those dividing cells? At about this time, some will form a column of cells the shape of a tiny tube. At one end of the tube, cells will develop into the spinal cord, while at the other end cells will develop to form the brain's two hemispheres. Most of these cells become neurons and at one end will have extensive branches–tens of thousands of them–called dendrites, which will act as antennae to receive input from other neurons. At its other end, each neuron will sprout a single long extension called an axon, which in turn will sprout at its tips tens of thousands of swellings called axon terminals.[1]

Within Sara's brain, these neurons will begin to send messages to each other via electrical signals that "leap" from the axon terminal of one neuron to a special docking site on the dendrite of another neuron, crossing a gap called a synaptic cleft. Eventually the brain will contain more than one hundred trillion such synaptic connections.

As Sara grows in her mother's uterus, her brain will continue to form as a sheet of cells called the cortex, which develops rapidly and extensively, so much so that it folds in on itself hundreds of times, giving the outer portion of the brain its cauliflower-like appearance. Other brain structures will develop beneath the cortex, but it is in the cortex that all of Sara's higher cognition will take place. This is ground zero for her future memory, attention, perceptual awareness, thought, language, and ability to make decisions. Sara's cortex in either hemisphere will further develop into five symmetrical lobes, which are recognized by landmarks on the surface of the brain and form in roughly the same place from one person to the next. Each lobe (roughly the size of an orange) will contain several billion neurons, which will exchange trillions of messages with each other, with neurons in the other hemisphere, and with deeper brain structures.

The cortex in the *frontal lobes,* just above and behind Sara's eyes, will develop to handle planning, execution, and control of movements. In the right hemisphere, her frontal lobe will be particularly tied to art and music appreciation as well as intonation in speech. In the left, it will be tied to logical thinking, concentration, and understanding symbols in reading and writing.

Near the ears, Sara's *temporal lobes* will develop to handle her sense of hearing and understanding language, among other things. Deep within each temporal lobe the hippocampus will form. It will be key to all aspects of memory.

Between her frontal and temporal lobes, the *parietal lobes* will develop to handle various sensory inputs from the skin. They will take charge of Sara's sense of touch, her ability to feel weight or motion, proprioception (the sense of knowing where parts of the body are at a given time, without having to look), and self-orientation. The parietal lobe on the right will help her with navigation, while the parietal lobe on the left will deal with calculation.

At the back of her brain, the *occipital lobes* will form. They will handle vision and will be integral to recognizing a familiar face or detecting a constellation of stars at night. Sara's occipital lobes will enable her to both see and make sense of what she sees.

Sara's forming brain will also include a *limbic lobe,* a collection of cortical areas in the frontal and temporal lobes, plus some deeper brain structures. One is the amygdala, an almond-shaped structure (actually there are two—one in each hemisphere), which is tied to emotions. Another component of the limbic lobe is the hypothalamus, which regulates hormones involved in a person's fight-or-flight response, eating, and metabolism, and is closely connected to the amygdala. Both the amygdala and hypothalamus are closely linked to the hippocampus.

The various parts of Sara's cortex will communicate with each other and with a dozen deep brain structures as well as the spinal cord via nerve bundles, which act as highways within the brain.

As Sara's brain develops, it is the conversation between neurons—those electrical signals passed and received—that keep them alive and connected. When they fire a signal across the synaptic gap, they strengthen each other, "wiring" them together and increasing their chances of survival. "Neurons that fire together, wire together," as neuroscientists like to say.

Neurons that don't fire and wire see their connections culled and shrivel away—which happens later in life through a phenomenon called disuse atrophy, or the "use it or lose it" principle. This process is critical during the initial stages of brain development but persists throughout life—a fact that is critical in our quest to grow the brain, as you'll soon read.

In the weeks before she's born, Sara's brain will continue to rapidly develop, so that by the time she enters the world it will be about 90 percent complete. Important areas needed for survival will have developed fully in order to ensure that she breathes, sleeps, and sucks when presented with anything that resembles a nipple.

The Phenomenal Hippocampus

You'll hear about the hippocampus over and over again in this book. And for good reason. The hippocampus, after all, is the gateway for new memories and essential for learning; as such, it is a major player in the quest for a bigger, stronger brain. Not only that, but the hippocampus is also the most malleable of brain regions. It is the first region to shrink with aging but also the quickest to grow in adulthood. And changes in its size bring noticeable changes in a person's memory and cognitive function.

That's not to say that your hippocampus works alone. The hippocampus is most instrumental in making new memories, but many other parts of your brain are also involved in creating and storing memory.

In simplistic terms, it might help to think of the hippocampus as a librarian. It processes all new information and decides what to keep and what to discard—tossing out those free advertising mailers, for example, while storing a copy of a front-page article of the *New York Times*. The good stuff—that which the hippocampus deems storage-worthy—is sent to various parts of the cortex for long-term storage. Information dubbed forgettable—like a phone number you're repeating in your head until you're able to type it into your phone—may be held for a short time but is then quickly tossed.

Although I'll often refer to this part of the brain in the singular, you actually have two hippocampi, one in each hemisphere of your brain. They look identical, and both bear a passing resemblance to a sea horse. (The name comes from the Greek words *hippos* for horse and *kampos* for sea monster.) Portions of the hippocampus are tied to spatial or emotional memory, while others handle a person's recall of a sequence of events or facts.

And yet, she's ready for survival in only the most basic sense. If a predator were to walk through the delivery room door, Sara couldn't get up and run to safety. She couldn't yet make the decisions—to gather food, work, protect herself from the elements—that would preserve her life as an adult.

Newborn Sara couldn't do any of these things, in part because her cerebellum—responsible for balance, coordination, and eye movement—hasn't fully developed, and in part because the connections between the neurons elsewhere in her brain have not yet fully formed, making it impossible for messages to be smoothly passed or functions to be fully coordinated. Her brain, at this point, is a little like a city under construction. And understanding how that city is built can shed light on how it can crumble—or grow—later in life.

Your Brain's Blueprint: CogniCity

Let's imagine Sara's developing brain as a collection of neighborhoods—these are the lobes you've just read about—connected by a complex series of highways, boulevards, and small roads. We'll call it CogniCity. Eventually, the highways in CogniCity will be smooth and perfectly paved, all the better to whisk cars—or messages, in this case—from point A to point B. When Sara's born, though, they're more like dirt roads connecting small towns. Travel is neither quick nor easy.

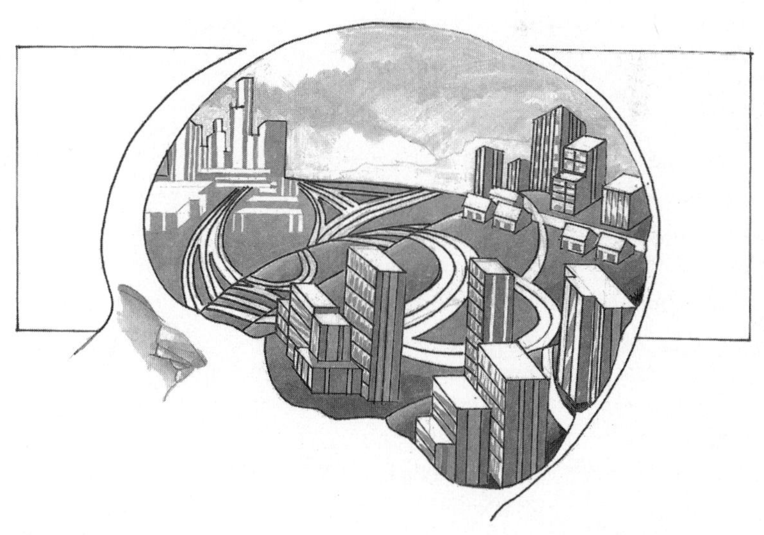

CogniCity

The brain can be thought of as a city made of neighborhoods of different sizes that are linked with each other by roads and highways.

That's, in part, because the nerve fibers that will one day become the brain's highways have yet to be covered with myelin, a fatty white substance that sheaths a neuron's axon in much the same way insulation covers an electrical wire. When myelin develops—in a process called myelination—it allows brain cells to pass electrical signals effectively and efficiently. Those messages travel down the axon on their way to the synapse as a series of signals, with each hopping further down in a process called saltatory conduction. Without myelin, this saltatory conduction requires a lot more energy to be expended by the neuron. And not only do unmyelinated axons expend more energy to do their work, they also do a less-than-stellar job of it. Some signals leak out of the unmyelinated axon, never to be delivered at all.

If myelination is not completed, some parts of Sara's brain can't communicate well with other parts, making complex movements harder, if not impossible, to pull off.

Beyond the roads, the neighborhoods have some development ahead of them as well. They will grow, almost exclusively, through the development of dendrites and axons. The two exceptions are the cerebellum and the hippocampus. The cerebellum isn't fully developed at birth and is one of the few parts of the brain that experience neurogenesis—the birth of new neurons—after a child is born. This happens throughout the first year of life. Incomplete development and incomplete myelination in the cerebellum mean a baby can't yet sit up or stand and can only clumsily grasp a bottle. The hippocampus, too, is not yet fully myelinated and will also experience neurogenesis, which, as we'll discuss throughout this book, lasts well beyond the first year and, in fact, into late life.

Still, these are the exceptions. In every other CogniCity neighborhood, future development is tied to the creation of synapses rather than the birth of new neurons.

As her brain's highways and neighborhoods develop, Sara's skills will begin to advance. While at birth she could only detect light, as synapses form and fiber bundles myelinate she'll be able to differentiate patterns and color in the things she sees. Before long, Sara will be able to smile (around the two-month mark), hold her head steady (four months), and sit without support (six months). By one year, she will most likely be able to stand and take a few steps.

In the years ahead, Sara will reach yet more milestones. By age three or four, the areas of the brain responsible for memory will have begun to mature: meaning her earliest memories will stem from this age, and not before. By age six or seven, she'll likely be able to entertain abstract thoughts. By eight or nine, she will almost certainly engage in more complex thinking.

As Sara grows, her brain will produce new connections—and cull those it doesn't need—shaping her brain in a process that continues until about the age of twenty, or perhaps even later.[2]

And It's Done . . . or Not

By the time Sara reaches her early twenties, her brain will be considered fully developed. And yet, that doesn't mean it's set in stone. As you'll soon read, Sara's brain will still have substantial capacity to grow and change.

Unfortunately, just as CogniCity develops, it can also deteriorate. When she enters her thirties, Sara's own CogniCity might already be showing subtle signs of decay—streetlamps flickering and flaming out, shutters hanging loose, small potholes forming in the city's oldest roads. As she ages, Sara may find her city slowly crumbling, losing synapses through lack of use, allowing her brain's fiber bundles to deteriorate and neurons to die from inadequate blood supply. Instead of a gleaming metropolis, her CogniCity might be a city in blight.

On MRIs, such brain aging shows up as a thinning of the cortex, which begins in middle age and continues into late life.[3] White matter fiber bundles—the network of connections between brain areas—also wither after age fifty, primarily due to reduced blood flow. The highly malleable hippocampus, meanwhile, is one of the first areas to shrink, and one of the hardest hit. It shrinks at a rate of 0.5 percent a year after the age of fifty, leading to poorer short-term memory. That's why as she ages Sara may experience subtle memory problems long before she begins to have trouble calculating a tip, for example.

When this all happens, and how much it affects a person, depends to a large extent on lifestyle choices. Sara wouldn't be too unusual if she began in her fifties to experience what we call age-associated memory impairment. This means she might book a hair appointment and then forget about it until the hairdresser calls, slightly annoyed. She might blank on the name of an acquaintance. Or forget the password she set up for an online account.

Such lapses might bring on acute fears of Alzheimer's disease for Sara, as it does for many people who have mild age-related cognitive changes. As you'll soon read, such problems may be accelerated by lifestyle choices or certain simple medical issues, resulting in a "brain fog" that can be easily treated and reversed.

At this stage, Sara's cognitive future would be at a bit of a crossroads: Take one path—with limited physical activity, untreated health problems, poor diet, and high stress—and she'd head toward a brain that looked like Mrs. Grey's. But if she followed the other path—if she grew her brain—Sara would likely be headed in a very different direction.

Your Brain Late in Life

As Sara ages, some amount of cognitive decline is inevitable. Typically, such decay unfolds in a pattern that looks remarkably like development in the young brain—only in reverse.

The parts of the brain that were first to get their full share of myelination and to fully mature are the last to deteriorate. Those that developed last, on the other hand, are the first to go. To understand why and how this happens, we can look at the hippocampus, which is one of the last areas of the brain to develop and myelinate (and actually is one of the only areas of the brain to never fully myelinate). With little to no myelin insulation, neurons in the hippocampus have to work overtime—for decades—in order to pass signals to other parts of the brain. Small wonder, then, that they are the first to wear out and die. This may be the main reason why Sara will likely experience memory lapses in her fifties and sixties, long before she has difficulty with the sensation of touch or her vision (which are among the first brain areas to develop). As Sara enters her seventies and eighties, other parts of her brain will begin to shrink in this last-in, first-out progression. Eventually, as she reaches (or even passes) her nineties, the parts of her brain involved in processing speed wither away. With enough time, she might even return to the simple cognitive functioning of a child.

This whole aging process, of course, happens to varying degrees and at varying speeds, depending in part on a person's life choices and overall health. There are, as you'll read throughout this book, a host of health and behavioral factors that can speed up—or slow down—the brain aging process. Most often, there are multiple factors at play. Alcohol abuse, traumatic brain injury, excess stress, sleep disorders, vitamin B12 deficiency, obesity, and cardiovascular disease (among other ills) can all destroy the brain's highways and neighborhoods.

A lifestyle with a focus on brain-boosting growth on the other hand, will help keep the brain young well into the last decade of life.

Studying the Brain

For years, much of how we tied particular actions to individual parts of the brain came from studying the brains of the dead or those with traumatic brain injuries. One of the most famous of such cases is that of a man named Phineas Gage, a then-twenty-five-year-old who in 1848 survived a railway accident in which a metal tamping rod shot through his skull. Doctors repaired the outward damage but there wasn't much they could do for his massive brain injury.

Quiet and pleasant before the injury, Gage—who recovered physically—was said to be volatile, rude, and impulsive after being hurt, although experts today disagree on just when those symptoms began to appear and how severe they really were. Still, Gage became a case study for understanding the functions of the frontal lobes. Since it was primarily his left frontal lobe that had been damaged in the accident, the change in Gage's behavior and demeanor offered insight into the role of the frontal lobes in decision making and impulse control.

Even more thoroughly studied than Gage was Henry Gustav Molaison, known in scientific circles as H. M. Born February 26, 1926, H. M. suffered a head injury in a bicycle accident at the age of nine. Perhaps from the accident—we don't know for sure—H. M. began to suffer frequent epileptic seizures. The seizures were so severe that in 1953 neurosurgeon William Beecher Scoville operated on him, removing most of his hippocampus on both sides of the brain.

The surgery was successful in reducing H. M.'s seizures. But it soon became apparent that the operation had robbed H. M. of something significant: his ability to acquire new information. Without his hippocampus, H. M. was unable to commit any new information to his long-term memory. So, while he could recall much of what had happened to him before his surgery, H. M. immediately forgot anything that occurred after it.

Over the years, H. M. became something of a poster child for memory and brain research. A succession of doctors tested and prodded him, all with the hopes of understanding his curious condition. Along the way, they gleaned valuable insight into the workings of the hippocampus. The fact that H. M. remembered events and people from early in his life told doctors that long-term memories weren't stored in the hippocampus. Similarly, the fact that he could still learn to perform a new physical task told doctors that his procedural memory was intact, and that this type of memory, therefore, probably wasn't stored in the hippocampus. His inability to recall new information, too, pointed to the hippocampus having a role in the making of new memories.

H. M. died in 2008. His brain, long a source of scientific interest, is still under study. In fact, at the Brain Observatory at University of California, San Diego, Project HM researchers are currently slicing into H. M.'s brain to see what new information it may yield.

Right about the time that H. M. went under the knife, another surgeon was compiling a treasure trove of data he'd collected while performing brain surgery. Dr. Wilder Penfield, a neurosurgeon at the Montreal Neurological Institute in Canada, had performed groundbreaking brain surgeries on epileptic patients throughout the 1940s. The surgeries involved deliberately killing neurons in an effort to reduce epileptic seizures. In order to determine which cells to destroy, Penfield delivered a small electrical shock to parts of the brain and then recorded the body's response. Patients were awake during the surgery—Penfield used local anesthesia—so depending on where the shock was delivered they might exhibit a twitch in the arm, for example, or suddenly recall a childhood memory. The technique served its intended purpose, helping Penfield determine which cells to target. But Penfield's records on how patients responded to shocks to various parts of the brain also allowed him to expand on the limited map of brain function that existed at the time.

As the field of neuroscience developed and grew, so too did the map. Studies on animals helped pave the way, as has the development of a variety

The (Much-Maligned) Middle-Age Brain

Even given that most people don't use their brains to their full potential, the middle-age brain probably gets a bit of an unnecessarily bad rap.

People in their fifties and sixties often report feeling mentally slower than they used to. Such slowing is typical and can be the result of a variety of factors, from normal aging, to medications they're taking, to lifestyle choices they're making. But while their brains probably aren't functioning as well as those of people twenty years younger, most people don't experience a *rapid* decline in processing speed, working memory, or learning in midlife. Absolute processing isn't typically a victim of normal aging either. A fifty-year-old can memorize a list of words just as well as a twenty-year-old, although perhaps not as quickly.

And as it slows down in some functions, the mature brain still has one advantage over the youthful noggin: experience. With every experience we have, we gain information that will factor into future decisions. As a result, by late middle age we tend to be more reasoned, less emotional, and better decision makers. If you don't believe that, you can ask my brothers. Twelve and fifteen years younger than me, more fit and with brains that are no doubt a little quicker than mine, they still lose to me in tennis pretty much every time we play. Why? Their hard hits and quicker reflexes are no match for my years of practice. Much to their chagrin, they have to admit that I know exactly how to place the ball just out of their reach. Experience trumps mental—and physical—speed.

of non-invasive technologies for humans, such as magnetic resonance imaging (MRI), functional magnetic resonance imaging (fMRI), electroencephalography (EEG), and positron-emission tomography (PET), among other tools. Such high-tech tools have allowed neuroscientists to peek beneath the human skull to pinpoint just which brain parts control which functions.

Today, advances in scanning technologies have continued to expand our ability to watch the brain in action, even right down to individual groups of nerve cells. Of course, while we've made mind-boggling progress in understanding how the brain works, there's still plenty of mystery left.

CHAPTER TWO

How to Grow a Brain

BY THE TIME I cradled Mrs. Grey's brain in my hands, I had already benefited from several years of medical education at Harvard and Johns Hopkins. I knew that a brain like Mrs. Grey's carried with it undeniable liabilities. Examining her mottled and shrunken cortex, I could see why Mrs. Grey's mental capacity had declined in her final years. With the destruction of neurons and synapses, the deterioration of fiber bundles, and the collapse of blood vessels, no doubt she would have struggled to think—to remember the names of loved ones, to navigate from her home to the grocery store, and to recall what she needed once she got there.

And yet, there were so many unanswered questions. Would Mrs. Grey's brain look different if she'd lived differently? How much different? What exactly would she have had to do?

Countless scientists have worked over the years to produce the research proving that brain shrinkage is tied to memory loss and other cognitive problems, and that certain actions—like the treatment of vascular diseases—may help slow such decline. It's an area I found so fascinating that, in 2009, several colleagues and I conducted an exhaustive review of the literature and documented the hows and whys in a paper published in the medical journal *Nature Reviews Neurology*.[1] That paper provided a concise explanation of all the factors that push us into cognitive decline late in life. In short, our research showed, late-life dementia results from a constellation of genetic and environmental factors—from certain Alzheimer's-related genes to obesity, diabetes, hypertension, head trauma, systemic illnesses, and obstructive sleep apnea—all of which work to shrink the brain.

It was a rather tidy summation of just what causes brain shrinkage. And it offered something of a recipe for helping the average person reduce his or her chances of meeting such a fate. The next question, of course, was "could we go beyond that?" It's one thing to slow the pace of brain aging but could we take it one step further and *grow* the brain—even a healthy brain? To find out, I dove into the research of the day, reviewing some two thousand medical journal articles. My focus was primarily on the hippocampus, since it is the brain's most malleable region.

You probably already know what I found, even without having read the resulting journal article, published in 2012 in *Nature Reviews Neurology*.[2] We can, of course, not only reverse the brain atrophy associated with aging but also expand the brain's size—even before shrinkage begins.

When an author submits an article for publication in a medical journal it is first sent to peers in the author's field. They provide the critical eyes that ensure published articles are scientifically sound. Often reviewers tear into such papers with critical zeal. It's part of the process. In this case, however, the reviewers were uniform in their praise: not one of the three neurology experts who reviewed the paper offered any major criticism.

I'll get you started in chapter 3 on your own plan to grow your brain, but first you need to know just how it's done. In short, brain growth occurs through four key avenues.[3]

The Core Four

Adding Synapses

We've long known that when you learn something new you create new synapses—a process called synaptogenesis—and that when you continue to use those synapses you strengthen them. But it's only in recent years that we've come to understand how significant the impact of such "brain training" can be and how quickly it can happen.

In fact, as you'll read in chapter 7, repeated stimulation of synapses can lead to measurable structural changes in the brain in as little as several weeks. Examine the brain of a person learning to juggle or play golf and you'll find the portions of the brain implicated for coordinated hand and eye movements—the cerebellum, parietal cortex, and frontal lobes—are stronger, and bigger, than in someone who never juggled or played golf. This is, in part, due to synaptogenesis. The same principle applies to ballet dancers, basketball players, mathematicians, violinists, cab drivers, or anyone else practicing or learning a new skill.[4]

Bolstering Highways

As you know, the brain's neighborhoods communicate with each other through a network of fiber bundles, which carry signals back and forth. When your hippocampus needs to communicate with your frontal lobes, for example, it sends a message via these pathways. How well they function depends on two things: how many synapses there are between the neurons that make up these fiber bundles and how well myelinated those neurons are. Adding synapses helps strengthen the connections, while preserving and nourishing a strong myelin coat on the axons through myelination helps to ensure messages zip speedily along.

Adding (and Aiding) Blood Vessels

The brain is a richly vascularized organ, fed by an intricate system of blood vessels that carry nutrients and oxygen to every cell. In fact, a third of the brain is made up of blood vessels and a full 20 percent of the heart's output goes to the brain, despite the fact that the brain only accounts for about 2 percent of a person's body weight.

The health and vibrancy of this network of blood vessels is crucial to the brain's ability to thrive and grow. If the network is clogged or shrunken—by blockages that slow or cut off the flow of blood—neurons can't thrive. Some die; some merely fail to grow. But growing the brain isn't just a matter of keeping existing blood vessels in good health. It also relies on actually increasing that network, by developing new blood vessel branches through angiogenesis. As you'll soon read, exercise has actually been shown to promote angiogenesis, which in turn brings more blood and oxygen to the one hundred billion cells in the brain.

Neurogenesis

We've long known that neurogenesis occurs in the developing brain. But only in recent years has the field of neuroscience offered up conclusive evidence that the human brain is capable of growing new cells even in adulthood, a concept that wasn't even taught to doctors in training during my medical school days in the 1990s.

That's not to say the foundations of such discovery were unknown. In fact, as early as the 1960s there was evidence that the adult brain is not fixed in size, that neurogenesis goes on well past childhood. It was a seemingly outrageous notion at the time and the scientific community largely dismissed it. That is until the late 1980s, when researchers conducting animal studies began to document the phenomenon in earnest. Before long they had proven that neurogenesis was not only happening, it was also closely tied to learning. A 1989 study of songbirds, for example, found that adult canaries grew new brain cells each year as they learned new songs in order to mate. Even so, as late as the early 1990s there was still a lot of skepticism in the field of neuroscience.

Enter Dr. Fred Gage, a professor in the Laboratory of Genetics at the Salk

Institute for Biological Studies who has studied neurogenesis extensively. For Gage and others, the 1990s would prove an opportune time to dispel the doubt, thanks to emerging tools that allowed them to stain, mark, and even count neurons in a way they never had before.

And dispel the doubt they did, offering up proof in 1997 that mice living in enriched environments—with social interaction, toys, and running wheels—had dramatically more hippocampal neurons than mice who lived in bare cages.[5] In fact, Gage's study showed that new neurons had grown in the brains of both groups of mice, but in the enriched group social interaction, toys, and running wheels seemed to help those fledgling neurons survive and mature. In that group, 90 percent of those new neurons thrived, compared to a normal survival rate of just 50 percent.

The results were fascinating, but the real clincher came when Gage and his team noted neurogenesis for the first time in humans.[6] That study labeled and counted new neurons born in the adult brains of patients with terminal cancers, who had agreed to donate their brains after they died. The human brain, Gage's study proved, wasn't fixed in adulthood after all. The news was a breakthrough in the way we think about the brain, crushing the myth that the brain only deteriorates with aging.

Since then, Gage and researchers around the world have helped to fill in the blanks about neurogenesis in the adult human brain. We now know that new neurons are born in the brain every day, but unless certain conditions exist (one of those conditions is the presence of the brain fertilizer BDNF, which you'll read about in a moment), they'll simply dissolve, never to grow into adult neurons. We know that neurogenesis happens throughout life but occurs to a lesser degree as we reach old age, giving us ever more reason to build up brain reserve in midlife. We know that other factors, such as sleep, may alter neurogenesis in the brain.

We also know that growth in the various areas of the cortex and hippocampus translates to better brain performance, as you'll read in part II of this book, and—perhaps most exciting of all—with simple behavior modifications it's possible to see remarkable change in the size of these brain areas in mere months.

In fact, as recently as 2012 evidence has emerged that some changes in the brain occur more rapidly than we knew—becoming evident after as little as two hours—and to such a large degree that they can be seen on MRI.[7]

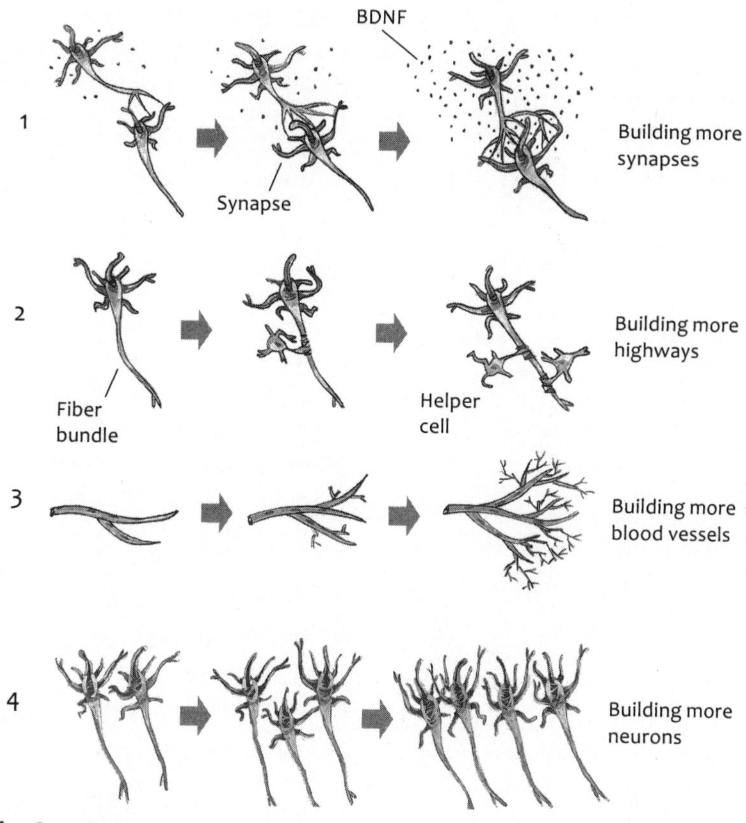

The Core Four

Brain growth can happen through lifestyle changes that are known to build synapses, bolster the brain's highways, nourish and grow its blood vessels, and promote the development of new neurons.

Fueling Growth

You've heard it a million times: eat brain food, exercise, sleep well, and you'll be rewarded with a "fit" brain. But what does that really mean? And what else can you do to enhance your brain's performance? As you'll discover, there is a wave of developing science behind a variety of techniques and practices that will put you on a path to grow your brain and boost its performance, both now and in the future.

In the past, efforts to boost brain performance have focused on slowing brain aging by preventing inflammation or limiting the risk of brain-damaging events such as a stroke. Those factors still matter when it comes to brain growth (your brain won't grow if it's shrinking due to harmful effects), but it's become increasingly clear that growing your brain really boils down to capitalizing on three mechanisms that are crucial to brain growth: oxygen, BDNF, and healthy brain wave activity. These mechanisms fuel the core four—adding synapses, supporting and growing a network of blood vessels, bolstering the brain's highways, and promoting neurogenesis. You'll read about them throughout this book, but here is the short story:

Oxygen

Unlike the cells in the rest of your body, neurons are constantly abuzz with activity, firing thousands of times a second, even when you're sleeping—a level of demand that requires tremendous oxygen flow. Oxygen keeps these neurons firing and is a critical ingredient in synaptogenesis and neurogenesis.

Optimal oxygen flow also aids in bolstering the brain's highways. Remember the myelin that covers a neuron's axon, allowing it to pass messages more quickly? That layer of insulation is actually kept intact with the aid of helper cells, which nourish and protect it. And what do those helper cells need to survive? Oxygen! Starve the helper cells and the myelin they're supporting degrades. The end result is a breakdown of communication between one neuron and the next. In our CogniCity example, you can imagine that happening along the communication highways (fiber bundles) that connect one brain part with another. Limit oxygen flow to the brain and you cause damage to both neighborhoods and highways.

You'll learn soon enough about factors that help and hinder oxygen flow to the brain. (Hint: They're the same types of things that keep us heart-healthy.) But for now it's enough to know that optimal oxygenation is key to boosting your brain.

BDNF

Brain-derived neurotrophic factor (BDNF) is a bit of an emerging rock star in the world of neuroscience. That's because, though we've long known of its existence, in recent years it has become increasingly clear that BDNF—a protein created by neurons—plays a critical role in brain health and growth.

For starters, BDNF is known to be a crucial ingredient in neurogenesis in the hippocampus. As you'll recall, the hippocampus constantly creates new neurons, perhaps as a sort of backup plan to replace those that are lost. If the fledgling neurons are needed, they grow to maturity; if not, they'll simply dissolve and dissipate. It is BDNF, though, that helps them grow. Think of it as a particularly effective fertilizer for the brain.

Scientists are in the midst of discovering the many ways we influence our brains' levels of BDNF, but there's already clear, compelling evidence that exercise, for one, helps boost our levels of BDNF and promotes neurogenesis.

The hippocampus isn't the only region that benefits from BDNF either. In the frontal lobes—the area of the brain responsible for abstract thinking and executive decision making—BDNF helps neurons heal and repair in order to stay in peak condition.

Healthy Brain Activity

The concept behind healthy brain activity—an exciting, emerging area of brain science—is simply that by training the brain to operate within an ideal frequency range, we can actually change the structure and function of the brain and enhance its performance.

To help you understand how it works, let me first give you a quick crash course in brain waves. The brain functions, as you know, by way of electrical signals. Those signals in groups of neurons, in turn, produce waves that oscillate at varying frequencies. And just as an orchestra is made up of a variety of instruments, each playing its own part, groups of neurons with a similar firing frequency produce a recognizable "song" or pattern. That song can be detected by sensors placed on the skull, using a technique called electroencephalography (EEG).

Each pattern of EEG is associated with a certain state of mind or level of brain function. For example, delta waves, the brain's slowest, indicate sluggish activity and are associated with sleep. Theta waves are also on the slow side and indicate either drowsiness or abnormally slow brain activity in a disease process, as seen in patients with traumatic brain injury. Middle-of-the-range alpha waves are associated with focus, calmness, and alertness, as well as creativity. Beta waves—faster than alpha—occur during heightened alertness and critical problem solving, though excessive amounts of beta are associated with anxiety, insomnia, and other problems.

Given the complexity of the brain, it's no surprise that one part can be motoring along at one frequency while another part can be recording a very different level of activity. If you're solving a math problem, for example, your left parietal lobe will likely be active, emitting beta waves as you puzzle out an answer, while the part of your brain responsible for orientation—the right parietal lobe—might be emitting slower alpha waves. That's perfectly normal. But just as you'd be alarmed if your heart raced all the time, a brain that's revving too high—or too low—without explanation is a sign that something's wrong.

Brain Wave Frequencies

Delta: 1 to 3 hertz

Theta: 4 to 7 hertz

Alpha: 8 to 12 hertz

Beta: greater than 12 hertz

We've long used EEG to measure brain activity—in epileptics, in stroke and coma patients, and for sleep studies—but in recent years a new discipline called quantitative EEG, or brain mapping, has emerged. Several independent groups of experts in the field have put together "normative" data from thousands of healthy individuals as well as hundreds of patients with various brain conditions, such as ADHD, autism, depression, anxiety, and OCD. In brain mapping, we use the available bank of normative data to look for signature patterns that tell us what may be going on in the brains of our patients. And we've discovered that we can tell quite a lot.

The normative data have clearly shown that a depressed patient, for example, exhibits excessively slow theta activity on the left side of the brain compared to the right. A person with ADHD typically exhibits slow activity in both frontal lobes, while a person with anxiety might exhibit too much beta activity on the right side of the brain.

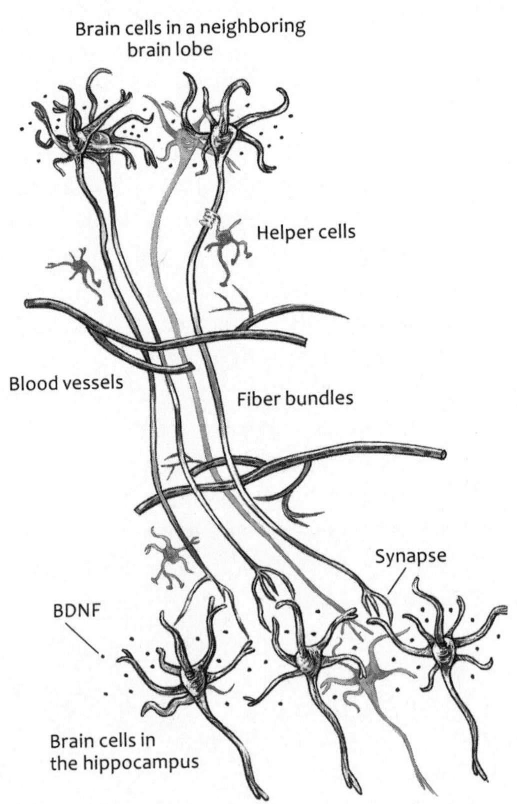

Before Brain-Boosting Efforts

A low density of neurons, synapses, BDNF, blood vessels, and fiber bundles in a person whose life is full of brain shrinkers and lacks any brain boosters.

A calm, relaxed brain, on the other hand, exhibits a pattern that is primarily in the alpha zone, with small amounts of theta or beta as well. This is healthy brain activity and, as you'll soon read, meditation, certain foods, exercise, and other activities promote it.

But how does healthy brain activity grow the brain and enhance cognitive performance? We don't yet have the extensive studies needed to fully understand how healthy brain activity changes the brain. But we do know that such activity as occurs during meditation is associated with decreased cortisol, which is a major brain shrinker, as you'll read in chapter 10. Healthy brain activity has also been tied to two members of the core four—angiogenesis and bolstering the brain's highways.

In other words, maintaining a calm and focused mindset actually makes new blood vessels grow and the network of connections between brain regions become stronger.

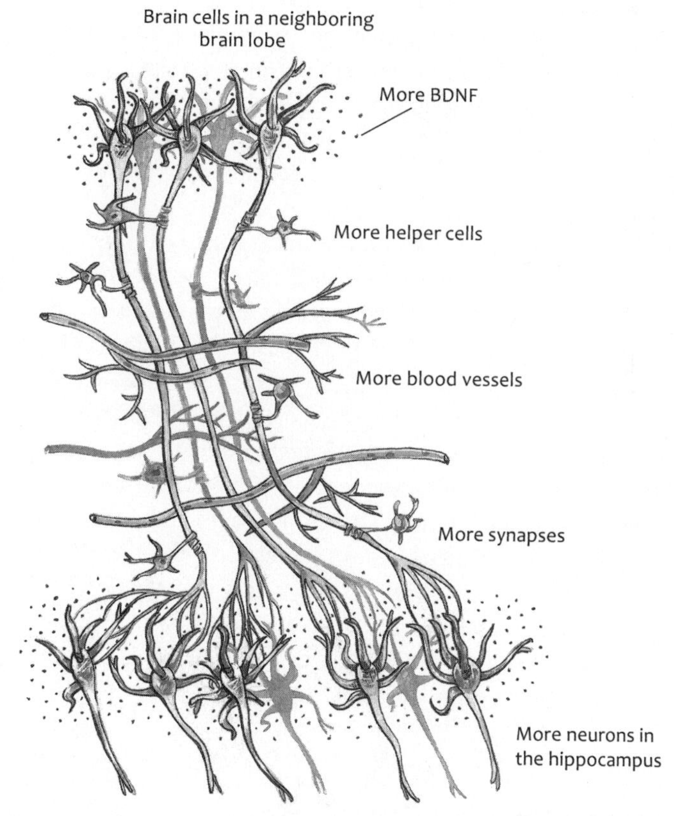

After Brain-Boosting Efforts

A high density of neurons, synapses, BDNF, blood vessels, and fiber bundles in a person whose life is full of brain growers and low on brain shrinkers.

Big Is a Boon

By now you know that the brain typically shrinks with age. And that the brain's incredible plasticity means you can not only offset that shrinkage but also enlarge a healthy brain. You know that such growth will help improve your clarity, memory, and creativity, and that it can also reduce your risk for late-life Alzheimer's disease. But why?

The answer lies in a concept neuroscientists call brain reserve. Brain reserve simply refers to the notion that promoting the core four—increasing the number of synapses in the brain, strengthening the connections across the brain, keeping the brain well nourished with rich blood flow, and growing new neurons—enhances brain performance now and results in a more resilient brain as you age.

You can think of it as a retirement fund, of sorts: pay into it early and often, and you'll have more to draw on in your later years. There's ample evidence for this. In one study, for example, people with large hippocampi were less likely to be demented late in life even if they had footprints of Alzheimer's disease pathology in their brains.[8]

One significant difference between financial planning and brain-reserve building is that a bigger brain translates to greater clarity, memory, and creativity not just in the future but now as well. Unlike a retirement fund, you don't have to wait until you clock out for the last time to benefit.

Keep in mind that even the smallest of the brain changes that we'll talk about in the pages to come—measuring mere fractions of a millimeter—reflect hundreds of thousands of neurons and tens of millions of synapses.

Of course, growth has its limits. As with muscles, once you reach the limit of growth in size, you'll still see increases in efficiency. After that, your brain will continue to become more efficient, although not larger, in much the same way your biceps muscles max out on size but continue to strengthen.

One good example to illustrate this may be the brain of bona fide brainiac Albert Einstein. Recent studies of Einstein's brain reveal it contained more intricate folds than the brain of the average person. Researchers are still puzzling out the meaning of this, but it's possibly an indication that Einstein's cortex had expanded in size and, in doing so, folded in on itself to a greater degree than most people experience.

Expand Your . . . Intelligence?

When I talk about having a bigger brain, I inevitably get eager nods and then a question along these lines: "But will that make me smarter?"

The answer is no. And yes. Actually, it depends on what you mean by smart.

If you Google the phrase "what is intelligence?" you will get more than six hundred million hits. Within those, you'll find a wide range of definitions for intelligence. That's because, despite much study and obvious interest, experts still don't agree on what, exactly, intelligence is. Still, most definitions include some or all of the following: the ability to engage in abstract thought, understand meaning in words, communicate with others, reason, learn, retain knowledge, plan, and problem solve. One Merriam-Webster definition puts it simply as "the act of understanding."

The definition I like to use is this: intelligence is the ability to function well in response to obstacles in life. In other words, it is your ability to figure out ways to be successful. This includes having the emotional intelligence—something IQ tests can't measure—required to interact effectively with others. It includes the ability to process information at a reasonable speed. And, yes, it includes the ability to remember things. But memory is just one component of intelligence. Being able to remember the elements on the periodic table might be helpful to success if your obstacle is a chemistry test, but it won't mean much when you need to come up with a new idea for the ad campaign that might earn you a promotion.

We're all born with certain innate abilities when it comes to intelligence, and those abilities vary greatly by person. One person may have a respectable ability to work with numbers, for example, while another may be terrible at math but can movingly convey meaning with words.

But—here's the beautiful thing—we *can* improve at least some of the components that make up our intelligence. Memory, creativity, mental agility (our ability to respond quickly or "connect the dots") all can be improved with a bigger brain.

We're only now beginning to understand how great that improvement can be. In early 2012, researchers at the University of Edinburgh published their findings that genetic factors account for about 24 percent of the change in intelligence that occurs between childhood and old age.[9] For the study, published in *Nature,* the research team examined the DNA and intelligence tests of two thousand people. The intelligence tests were conducted when the subjects were children, and again when they were in old age. The largest impact on intelligence, researchers found, seemed to come from environmental rather than genetic factors—what we do rather than what we are given.

To get back to the question—does having a bigger, stronger brain mean you'll be smarter—I firmly believe it does and that you'll perform better on tests of intelligence, be they real world or paper and pencil.

Brain Reserve in Action

You know when your brain isn't functioning well. And when it is. In peak shape, your brain can perform with incredible efficiency and effectiveness—recalling memories quickly and accurately, solving problems, and thinking creatively. Each part of the brain chugs away independently while at the same time communicating and coordinating with other parts of the brain in a way that's as beautiful as it is mysterious.

A brain operating at its worst, on the other hand, isn't a pretty sight. It makes poor decisions, is stumped by challenges, and lacks creativity. The results are far-reaching. Your brain, after all, is behind every function your body engages in. From each breath you take, to the words you speak, to the complex decisions you make each day. Your brain is at work when it tells your arm to lift your toothbrush to your mouth; it's at work when you choose to pour milk, instead of orange juice, on your cereal; it's at work when you decide to hold your tongue rather than blurt out a snarky comment about your boss's work habits.

At every stage of life, how well your brain functions will affect how well you function—at home, in relationships, at work, at play. There's a circular aspect at work, too: a poorly functioning brain might make decisions—on what to eat, how much to sleep, how active to be, or how much alcohol to consume—that reduce your ability to make good decisions. The result: more bad decisions.

If you're not yet convinced, consider these two drastically different variations of a day in the life of Martina, a mid-level manager working for a large insurance company.

In the first scenario, we have a seriously brain-drained Martina, whose unhealthy lifestyle has left her chronically sleep deprived, overweight, short on physical activity, and perpetually trying to catch up.

On this day, Martina oversleeps by a full thirty minutes, courtesy of her insomnia and anxiety over a work problem, which have kept her awake on and off throughout the night. Groggy, frustrated, and short on time, Martina rushes her children to the bus stop and then races to her car, choosing by default to skip breakfast at home in favor of a muffin from her favorite coffee shop. Stopping for coffee puts her a little further behind schedule. Frazzled, she decides to take a "shortcut" to work, forgetting that road construction is under way and will make her even later. Gripping the steering wheel tightly, she silently seethes, blaming her predicament on her husband's failure to help around the house. She would have been on time, she thinks, if he'd taken charge of prepping the kids for school.

By the time she gets to work, she's annoyed, stressed, and feeling already behind schedule. She sits down at her desk and opens an e-mail from a coworker who tends to butt heads with others. Irritated, she forwards it to a friend with a sarcastic comment about the original sender. But instead of hitting forward, Martina has actually hit reply. She realizes her mistake immediately and spends the rest of the morning trying to undo the damage. Feeling distracted and disappointed in herself, Martina attempts to prepare for a presentation she must give but finds herself lacking the creative juice the task requires. Hoping to complete it anyway, she skips lunch and spends the hour writing and rewriting the same four paragraphs.

The rest of the day doesn't go much better. At 3:00 P.M. she fields a teary call from her daughter, who'd rushed out of the house without the cleats she needed for after-school soccer. Feeling guilty, Martina dashes home, then to school to deliver the shoes, and then back to work. Hours later, Martina is rushing again, this time to make it home in time to feed the family. On the way, she picks up pizza and a six-pack of sugary soft drinks.

Ahead of her is another two hours of work, which she completes on her laptop while lying in bed. Then she's in for another night of interrupted sleep. The next day will unfold much like this one, with Martina feeling perpetually behind the ball, mentally fuzzy, and uninspired.

Now, let's imagine Martina under a different set of circumstances, leading a life that has boosted her brain. Martina sleeps soundly and awakens on time, feeling refreshed. In fact, she has the time and energy to cook a healthy breakfast for herself and her children. Before she leaves for work, Martina remembers to pack her daughter's cleats and then walks her children to the bus stop. On the way, she enjoys their childish chatter and snags a hug and kiss from each.

In the car, Martina remembers that road construction has made her usual shortcut anything but short, so she opts for another route instead. She arrives at work on time and ready to tackle the day's challenges. Opening an e-mail from that annoying colleague, she keeps her frustration in check and responds with a professional reply. In her interactions with colleagues, she's upbeat and savvy. She remembers to ask a coworker about his father's recent surgery and to call a client by the nickname he prefers. Her positive energy won't go unnoticed by her boss, nor will her creativity and problem-solving abilities, which will help propel her up the corporate ladder.

At lunch, she heads out of the office for a thirty-minute power walk and returns feeling energized. In the midafternoon, she takes a break for ten minutes of breathing exercises, followed by a cup of herbal tea. She finds herself feeling revived at 4:00 P.M. and sends a flirty text to her husband. Glancing at her planner, Martina notes that she has to leave the office at 5:30 P.M., so she can get home in time to cook a brain-healthy meal and then arrive on time for a monthly book club meeting. She powers through a stack of paperwork. At 5:32 P.M., she slips the key into her car's ignition.

At home, Martina's husband and kids share the details of their day over a meal of salmon, spinach, and brown rice. Then Martina's off to her book club meeting. She'll be home in time to cap off the evening with her husband and will be fast asleep by 10:00 P.M.

Of course, even the fittest of brains can fumble on occasion, so the Martina of the second scenario isn't in for a life that's completely error-proof. But she's clearly running on a full eight cylinders, performing at her cognitive peak with results most people would envy.

How does she do it? Martina's enhanced cognitive performance is no accident. Achieving it took a concerted effort—boosting BDNF, increasing oxygen flow, and promoting healthy brain activity in ways that fueled the core four of brain growth.

How can you do it? You're about to find out.

Epigenetics: Nurture Alters Nature

You probably know someone who is remarkably good at remembering trivia. Or who can juggle numbers with ease, and without a calculator. Or who can grasp the meaning of complex notions well before his or her peers.

Some people are better at understanding and learning how to play a musical instrument; some are better at athletic pursuits; some are better at expressing themselves through language. Some are quick thinkers; others are not.

In part, these skills are based on the genes you inherited from your parents. That's nature. And it's important. Nature, which accounts for the color of your eyes or your blood type or whether you're male or female, also accounts for some of the basic cognitive abilities you're born with.

But perhaps even more important is nurture, or your environment, which affects how your genes express themselves—a phenomenon called epigenetics.

To understand how epigenetics works, imagine the brain of a violin prodigy. If we could examine it, we'd likely see that the part of her brain that handles music is more developed than it would be in the brain of her non-musical peer. She was likely born with an affinity for understanding music—that's nature—but as she developed her musical skills with regular and intensive practice, the parts of the brain responsible for playing the violin also developed and grew more (hello, nurture) than if she never had practiced.

Our violinist can put nurture to work growing other parts of her brain, too. For example, practicing memorization skills will make her better able to recall the names of composers, or her friends' phone numbers, or even her credit card number. Even though she was not born with any particular genes related to exceptional memory, she can still become a memory champion.

CHAPTER THREE

Creating Your Twelve-Week Plan

IF YOU'VE PICKED UP this book, you're probably on a mission. You want to think more clearly and boost your memory and creativity, now and in the future. To do so, you'll need to look at your brain in a new light—as a vital organ that needs the same care and consideration you give your heart, your skin, even your teeth! Your ability to grow it rests on maximizing factors that grow the brain (coming up in part II) and minimizing factors that shrink the brain (coming up in part IV).

I see patients much like you at my Brain Center. Some are young and relatively cognitively healthy but in search of ways to sharpen their thinking. Others are older and hoping to ward off Alzheimer's disease. Still others are already struggling with persistent memory problems or mental slowness or with the cognitive effects of a serious condition, such as a traumatic brain injury, stroke, or depression.

Almost all—even the relatively healthy ones—are surprised to discover the extent to which they're shrinking their brains through their lifestyle choices. They have no idea of the true cost of the donut they had for breakfast, or the five hours of sleep they consistently survive on, or the high-stress career they've chosen. They'll soon learn all about such brain shrinkers. And they'll also learn exactly how to *grow* their brains—by adopting prescribed habits that range from walking five days a week, to daily mindfulness, to enriching their diet with brain growers, such as DHA or resveratrol.

Many are eager to enroll in my twelve-week brain fitness program. Once I've explained to them the dynamics of growing their brains, I hand patients a pocket-size booklet they'll use to track their progress. We call it a "passport." It's a fitting reference because this passport not only traces a person's journey to a bigger brain, but it is also his or her entrée to a new world of brain fitness. This book is your passport. Through it you will create your own customized twelve-week plan, outlining in detail the exact steps you'll need to take each week to arrive at your destination—a bigger brain and the enhanced memory, clarity, and creativity that come with it.

Each step of the plan is based on my experience in seeing thousands of patients and on my in-depth research into the latest discoveries in neuroscience. The goal is creating a thicker, denser, healthier cortex and hippocampus—by boosting BDNF, increasing oxygen flow to the brain, and maximizing healthy brain activity.

I'll begin where I start with all my patients: getting them—and now you—into the mind-set to sculpt a bigger, stronger brain in twelve weeks.

Get in the Zone (and Aim High!)

From my very early years, my father assured me that I would one day be successful. In fact, he sort of demanded it. "You are going to be a prominent scientist, a professor who is known worldwide. You will speak seven languages. You will win the Nobel Prize," he would boldly declare.

These were lofty goals, but he said them as if he believed them. And he made believing them a part of my life. When I was a kindergartner, he would lead me out of our home to his sky-blue Volkswagen Beetle, open the door for me and, with a sweeping arc of his arm, usher me into the backseat. "After you, Dr. Fotuhi," he would say, signaling to me the future I might expect if I achieved my goals. I would clamber into the backseat and he would gently draw a comb through my hair, straighten my collar, and brush imaginary lint off my imaginary suit jacket. In that moment, I *was* on top of the world, at least in my mind's eye.

The Nobel Prize is still a part of my dreams. But I did go on to become a doctor, speak five languages, and lead the sort of life my father had envisioned for me. What got me there? Was it luck? Maybe. But a large part of achieving those objectives was setting clear and specific goals and having the passion and motivation to propel myself ever forward toward them.

You can do the same.

First, you'll need to set goals. Then you need to become excited about them. Think about why you want to grow your brain. Do you want to be more effective at work so you can get a promotion or advance your career? Do you want to stop forgetting names so you feel more comfortable in social situations? Or think more clearly so you're more effective in juggling your family's schedule? Are you worried about Alzheimer's disease and want to ensure you're mentally sharp enough to beat your grandchildren at their computer games? Whatever your goals may be, they're far more likely to be achieved with a bigger brain. I assure you that, once you complete your twelve-week plan, you *will* have better memory, clarity, and creativity.

With that in mind, write down three chief goals in your life (be specific):

1. ______________________________

2. ______________________________

3. ______________________________

Now it's time to find that passion. Look two years down the road. Imagine that every day you are boosting blood flow to your brain, churning out BDNF, and chugging along in optimal brain wave territory. Picture yourself greeting each day with energy and tackling your tasks with more creativity, clearer thinking, and a better memory. You've changed your lifestyle—and boosted your brain. Picture yourself remembering everyone's name at the office Christmas party. Or finally feeling fully on top of your children's ever-changing schedules. Or winning your dream job. Imagine yourself at eighty taking a college course, just for the fun of it. Or at seventy writing a book, just because you can. Picture yourself healthy and strong and thinking like a man or woman ten years younger.

On the following lines, I want you to write several sentences vividly describing the life you've just envisioned. Write it as if you've already achieved those goals.

Example: I completed this program last month. Now I can easily remember the name of every person I meet.

____

____

____

____

____

It sounds pretty appealing, doesn't it? This is the you of the (not too distant) future. You're ready now to make it a reality.

Your Brain Fitness Assessment

Before you implement any changes, you need to understand just where you stand in terms of all the factors that affect brain health. If you walked into my

Brain Center today with concerns about your memory or cognitive function, there are some standard steps I would take:

1. I'd review medical records from your other doctors to learn about any conditions you've had in the past.
2. I'd listen to your concerns and obtain detailed information about when symptoms began, how severe they are, and if friends or family have noticed them.
3. I'd ask questions about your health, from head to toe.
4. I'd perform a physical and neurological exam, with special emphasis on checking your memory, attention, and concentration.
5. I'd formulate a potential diagnosis for you and order one or more of the following tests, as needed:

   –a brain MRI, to measure the exact size of your hippocampus;

   –a sleep study, to check for sleep apnea;

   –an actigraphy, a five-day test that measures insomnia at night and activity during the day;

   –an ultrasound of the blood vessels in your neck, to check for signs of blockages in the blood flow to the brain;

   –a cardiopulmonary exercise test, to measure your "VO_2 max" (a measurement of stamina and endurance);

   –a complete cognitive evaluation, with paper-and-pencil tests to measure the function of each area of the cortex and hippocampus;

   –a quantitative EEG brain mapping, to check brain wave frequency and look for signs of anxiety, depression, insomnia, or ADHD; and

   –blood tests, to assess levels of vitamins B12 and D, ferritin, cholesterol, and CRP (C-reactive protein), or possibly other tests if your history hints at kidney or liver problems, low testosterone, Lyme disease, or other conditions.
6. I'd review the results of your tests with you and inform you of your brain fitness status.
7. I'd set you up to meet with our brain coach and start you on an individualized three-month brain fitness program.

Since you're not walking into my office and instead are using this book, I can't perform these health checks. You will need to see your doctor for a checkup and obtain a set of basic tests. You will also need to discuss your other health issues and lifestyle habits, as you will learn in a moment.

Know Your Numbers

There are certain overall health measurements that also reveal information about your brain's health and warrant knowing. One is *blood cholesterol,* a vital stat to keep on your radar, especially since high levels of blood cholesterol can cause damage without producing symptoms. As you'll read in chapters 11 and 12, high blood cholesterol and *hypertension,* or high blood pressure, are two warning signs that your brain isn't getting optimal oxygen flow. They're also risk factors for heart attack and stroke, the ultimate oxygen starvers.

A healthy blood cholesterol profile, according to the American Heart Association[1] is:

> *Total Blood Cholesterol:* less than 200 milligrams per deciliter (200 to 239 is borderline high; 240 is high).
>
> *LDL:* less than 100 mg/dL is optimal (130 to 159 is borderline high; 160 or more is high; more than 190 is very high). Patients with diabetes should have an LDL less than 70.
>
> *HDL:* more than 60 mg/dL is best (for men, 40 to 49 is average; for women, 50 to 59 is average; less than 40 mg/dL for men and 50 mg/dL for women is low and considered a major risk factor for heart disease).

If your total blood cholesterol is above 200, your LDL is above 160, and your HDL is below 40 for men or 50 for women, you may need an annual cholesterol test.

You should also know your *blood pressure* reading. Ideally, you'll do this as part of a regular visit to your physician, annually at least, but you can also keep an eye on your blood pressure by using blood pressure machines available in many pharmacies or buying one for home use.

A normal blood pressure is at or below 120 over 80 (120/80). If your blood pressure falls between 120–139 over 80–89, you may be diagnosed as prehypertensive. If it is persistently greater than 140/90, you may be diagnosed as having hypertension.

You should also know your blood levels of:

- *Thyroid:* One measure of thyroid function for screening purposes is your level of thyroid-stimulating hormone (TSH). Normal TSH levels are between 0.4 and 4.0 mIU/L (milli-international units per liter). If your doctor is concerned that you may have a thyroid problem, he or she might order a full thyroid panel.
- *Vitamin B12:* The standard "normal" amount for B12 is above 200 pg/mL (picograms per milliliter), but knowing the importance of B12 for brain function, I always help my patients maintain levels

above 500 pg/mL. For example, you might be told you have normal levels of B12 (if your level is above 300 pg/mL) but I'd recommend taking B12 supplements until your level reaches 500 pg/mL.

- *Vitamin D:* Low vitamin D is a common problem and is associated with many neurological conditions as well as fatigue, memory loss, and dementia. The standard "normal" amount for vitamin D is above 40 nmo/L (nanomoles per liter), but given vitamin D's importance in brain health I (along with most clinicians) favor levels above 60 nmo/L.
- *Hemoglobin A1c:* Glycosylated hemoglobin A1c, or HgA1c, is a measurement used to detect diabetes. People who have HgA1c greater than 6.5 percent have diabetes. A level of 5.5 percent to 6.5 percent signifies pre-diabetes, and a level below 5.5 percent is normal.
- *Fasting Blood Glucose:* A normal fasting blood glucose level is less than 100 mg/dL. A level of 100 to 125 mg/dL indicates pre-diabetes and 126 mg/dL or greater indicates diabetes.
- *C-Reactive Protein:* CRP is a measure of cardiovascular risk factor and it should be less than 1 mg/dL. Between 1 and 3 mg/dL is considered intermediate risk, and greater than 3 mg/dL is considered high risk. Sometimes elevated CRP is seen in conditions other than cardiovascular disease, such as rheumatoid arthritis.
- *Ferritin:* Ferritin level is a measure of iron storage. A normal ferritin level for men is between 12 and 300 ng/mL (nanograms per milliliter); for women, it is between 12 and 150 ng/mL. A level below 12 ng/mL is associated with iron deficiency and anemia, and a high ferritin level is associated with alcoholic liver disease.
- *Testosterone:* In men, the normal range for testosterone is between 300 and 1,000 ng/dL (nanograms per deciliter). Low testosterone is associated with reduced libido, erectile dysfunction, fatigue, lack of energy, and difficulty concentrating. And it's not as uncommon as you might think. I see many patients who have low testosterone and don't realize it.

In addition to your blood levels, you should be aware of your weight status. A good measure of healthy weight is the waist-to-height ratio (WHtR), which is your waist circumference in inches divided by your height in inches. A healthy WHtR is under 0.5.

Now it's time to see where you stand. In the following table, list your health measurements. If they're not in the normal ranges, talk to your doctor about treatment. If you don't yet know your measurements, be sure to get them and fill in this chart when you do. You can also come back to this page at the end of your twelve-week program and fill in your measurements at that time (in the three-month column).

My Health Measurements

	Current	After three months	Ideal
LDL cholesterol			Less than 100 mg/dL
HDL cholesterol			More than 60 mg/dL
Blood pressure			At or below 120/80
Thyroid (TSH)			Between 0.4 and 4.0 mIU/L
Vitamin B12			More than 500 pg/mL
Vitamin D			More than 60 nmo/L
Hemoglobin A1c			Less than 5.5 percent
Fasting blood glucose			Less than 100 mg/dL
C-reactive protein (CRP)			Less than 1 mg/dL
Ferritin			Between 12 and 300 ng/mL for men; between 12 and 150 ng/mL for women
Testosterone (men)			Between 300 and 1,000 ng/dL
Waist-to-height ratio (WHtR)			Waist should be less than half your height in inches (less than 0.5)

Get a Baseline

Your next step toward a bigger brain is to use the following calculator to determine your baseline "Fotuhi Brain Fitness Score." This is a rough estimate of your current brain health status based on modifiable factors in your life, intended merely as a big-picture view of your brain function at this time. You'll complete this chart again at the end of your twelve-week brain fitness program. By then your score will have risen significantly.

Brain Fitness Calculator

		Score
LDL cholesterol	Give yourself a 1 if your LDL is high (more than 160 mg/dL), a 3 if your LDL is low (less than 100 mg/dL), or a 2 if it falls somewhere in between.	
HDL cholesterol	Give yourself a 1 if your HDL is low (less than 40), a 3 if your HDL is high (more than 60), or a 2 if it falls somewhere in between.	
Blood pressure	Give yourself a 1 if your average daily blood pressure reading is regularly 140/90 or worse, a 3 if it is regularly 120/80 or better, or a 2 if it falls somewhere in between.	
Vitamin B12	Give yourself a 1 if your B12 is less than 200 pg/mL, a 3 if it is more than 500 pg/mL, or a 2 if it falls somewhere in between.	
Vitamin D	Give yourself a 1 if your vitamin D is less than 40 nmo/L, a 3 if it's more than 60 nmo/L, or a 2 if it falls somewhere in between.	
Sleep quantity	Give yourself a 1 if you sleep fewer than six hours each night, have trouble falling asleep or staying asleep, or wake up extremely early. Give yourself a 3 if you fall asleep easily, get seven to eight hours of sleep each night, and wake up feeling rested. Give yourself a 2 if your sleep habits place you somewhere in between.	
Sleep quality	Give yourself a 1 if you snore loudly, feel very sleepy during the day, have no energy, and feel mentally foggy; a 3 if you never snore and do not have sleepiness during the day; or a 2 if you suspect you're somewhere in between.	
Weight	Give yourself a 1 if your waist-to-height ratio (WHtR) is 0.7 or more, a 3 if it's 0.5 or less, or a 2 if it falls somewhere in between.	
Activity level	Give yourself a 1 if you have a sedentary, "couch potato" lifestyle, a 3 if you get forty-five minutes of vigorous exercise five days a week, or a 2 if you're somewhere in between.	
Brain safety	Give yourself a 1 if you regularly engage in activities that can damage the brain (not wearing a helmet when cycling, skiing, horseback riding, or snowboarding; failing to wear a seat belt; boxing or practicing mixed martial arts), a 3 if you always protect your head or don't engage in brain-risky behaviors, or a 2 if you believe you fall somewhere in between.	

		Score
Social engagement	Give yourself a 1 if you lead a very isolated life and consciously avoid interacting with others, a 3 if you're a social butterfly and enjoy meeting new people, or a 2 if your social activity level falls somewhere in between.	
Alcohol use	Give yourself a 1 if you drink three or more glasses of alcohol daily or if you binge drink on weekends. Give yourself a 3 if you drink only socially or have on average one serving of alcohol a day (two for men). Give yourself a 2 if you fall somewhere in between.	
Food quantity	Give yourself a 1 if you overindulge, get second servings, and snack often; a 3 if you eat healthy portions with no more than one or two snacks a day; or a 2 if you fall somewhere in between.	
Food quality	Give yourself a 1 if your diet consists of fatty burgers, pizza, salty french fries, and sugary sodas; a 3 if your plate is usually like a rainbow, with a good mix of high-protein, low-carbohydrate, heart-healthy ingredients, and you eat fruits, vegetables, and nuts for snacks; or a 2 if your diet quality is somewhere in between.	
DHA	Give yourself a 1 if you never consume fish and do not take a DHA supplement, a 3 if you eat fish twice a week and take 1,000 milligrams of DHA daily, or a 2 if your DHA consumption falls somewhere in between.	
Mood	Give yourself a 1 if you are regularly sad, depressed, or irritable; a 3 if you are almost always cheerful, upbeat, and agreeable; or a 2 if your mood is somewhere in between.	
Mindfulness	Give yourself a 1 if you are stressed on a daily basis, a 3 if you rarely feel stressed and are often in the "alpha zone," or a 2 if your state of mind falls somewhere in between.	
Attitude	Give yourself a 1 if you are a pessimist, a 3 if you are an optimist, or a 2 if your attitude is somewhere in between.	
Memory stimulation	Give yourself a 1 if you never try to remember names or any other information; a 3 if you try to memorize the name of every person you meet and practice memorizing your grocery list, facts, or even poetry; or a 2 if your memory stimulation falls somewhere in between.	
Sense of curiosity	Give yourself a 1 if you rarely attempt to learn new skills, hobbies, or activities; a 3 if you try daily to learn new skills, hobbies, or activities; or a 2 if your sense of curiosity puts you somewhere in between.	
Total		

Use your final Fotuhi Brain Fitness Score to determine your current brain health status.

Green: 50 to 60 Good news! You're starting your journey with a relatively fit brain. There's always room for improvement, though, so your twelve-week plan will focus on fine-tuning your lifestyle choices to grow your brain.

Yellow: 35 to 49 You're starting your journey in relatively good brain health but with some problem areas. You'll likely need to identify key weaknesses and devote a strong effort to strengthening them, while maintaining brain-healthy habits in other areas.

Red: 20 to 34 Your brain is in need of some TLC. Your goal over the next twelve weeks will be to improve your scores across the board. This may involve some major lifestyle changes. As you read the chapters that follow, think about which lifestyle changes you can realistically make and which will have the biggest impact for the effort. Committing to regular exercise, as you'll read, may provide greater benefits than memorizing your grocery list. For patients who have many changes to make, or who have physical limitations, I advise tackling one or two meaningful changes at a time and adding in other changes as the weeks progress. Please work closely with your doctor or other health care professional.

Warning Card

This assessment is only a reflection of your current lifestyle and health conditions. Just as years of big spending might drain your retirement funds long before you reach old age, years of poor health can shrink your brain.

If you've had brain-shrinking health conditions or lifestyle factors in the past, you might already have decreased your brain reserve. Don't panic. Growing your brain over the next twelve weeks—and into the future—will help you more fully fund your reserve.

We can't tell, without a detailed study, just how much effect those brain drainers have had. So, if you've had past behaviors that are currently under control or are not evident but might have taken a toll on your brain, give yourself a "warning card." Consider this a reminder that your brain may be in need of all the brain-building efforts you can muster.

High priority factors that warrant a warning card include a history of the following conditions:

- Diabetes
- High blood pressure
- Sleep apnea

- Heart disease
- Stroke
- Smoking
- Insomnia
- Excess weight or obesity
- Sedentary lifestyle
- Extremely poor diet
- Head trauma

Digging Deeper into the Brain–Body Connection

When patients talk to me about their memory or brain health concerns, it's often clear that they think of their own brains as mysterious black boxes, perched upon their shoulders and orchestrating their thoughts and actions in isolation from the rest of their bodies.

The brain is certainly complex, and awe-inspiring, but while it stands alone in one sense as the body's command center, it is also an integral part of a network of organs, all of which affect—and are affected by—each other. Often my patients have no idea just how interrelated the health of their various organs is and have never really considered how far-reaching certain health problems can be.

This interconnectedness is especially important given that the human brain is in a constant state of flux, changing from hour to hour and day to day.[2] Health conditions that fog the brain, limit blood flow, or change hormone levels, even for mere hours, can leave their mark on brain structure, size, and performance.

In chapter 8 you'll get started on implementing a twelve-week plan to grow your brain. But before you move ahead, it's important to consider—and begin to treat—medical conditions that may impede your efforts.

Key Health Conditions That Affect the Brain

To build a bigger, stronger brain you need a strong, healthy body. Here are some common health problems that impair brain function, or even reduce brain size:

Hearing Loss Hearing, of course, provides you with auditory input, giving the brain a rich source of information to respond to and remember. Eliminate or reduce the input and you have a problem that goes far beyond just the inconvenience of constantly asking others to repeat themselves. Hearing loss, a common problem associated with aging, can actually hasten cognitive decline. As hearing fails, the elderly can find themselves increasingly isolated and less likely to use their cognitive skills. As a result, they experience "disuse atrophy"—shrinkage in the parts of the brain they've stopped using. But it's important to note that hearing problems aren't only for the elderly. There's recent evidence that mild hearing loss in the young is on the rise.[3] Experts don't know for sure why this is, but some suggest it may be linked to listening to loud music or to the use of earbuds, a type of headphone that sits snugly in the entrance to the ear canal.

Vision Loss Just as the ears provide input you need for stimulating your brain and cognitive function, so too do the eyes. With aging, the risk of developing glaucoma and cataracts rises, although vision problems can occur at any time in life. Excessive computer use, or prolonged staring at any electronic device, can also cause eyestrain or blurred vision.[4] Not only can such symptoms be unpleasant, but they can also reduce your cognitive function by making it hard to absorb information. Fortunately, vision problems can often be corrected with glasses, contact lenses, or surgery. Symptoms from computer use usually resolve once you stop using your computer, but they can also be avoided by limiting glare, adjusting your distance from your computer screen, and correcting minor vision problems.

Insomnia Insomnia is a critical sleep disorder that can reduce brain function, as you'll read in chapter 9. Often people ignore their insomnia, considering it just one more fact of life to be dealt with. They don't realize that insufficient sleep could be shrinking their brains and crimping their cognitive performance. What's more, they fail to realize that insomnia can be treated. In addition to many herbal supplements or various medications, meditation and neurofeedback have been shown to reduce sleep problems, either directly or by reducing anxiety.

Snoring Snoring can be benign, but it can also be a sign of obstructive sleep apnea (OSA), a sleep disorder that starves the brain of oxygen, seriously shrinking the cortex and hippocampus. Unfortunately, it's often a problem that goes ignored or is even laughed about. (Patients sometimes gleefully report they get the master bedroom to themselves because their log sawing has driven their spouses to the guest room.) But treatment for OSA, as you'll read in chapter 9, can be highly effective, reversing the damage done in the brain.

Medications There are a host of medications, especially in high doses, that can temporarily affect cognitive function. These include anxiety medications, such as Valium; sleeping medications, such as Ambien; narcotic pain medications, such as Percocet; and older antidepressant medications, such as amitriptyline. I often favor minimizing my patients' medications as much as possible. Medications that dull your senses and prevent you from absorbing information can lead to disuse atrophy–shrinkage of the brain–in the long run.

Smoking In addition to boosting your cancer and other health risks, smoking raises your risk for brain shrinkage and cognitive decline.

Lung Problems Uncontrolled asthma, chronic obstructive pulmonary disease (COPD), or emphysema make it harder to exercise strenuously, limiting oxygen flow to the brain and putting the brakes on brain growth.

Gum Disease Studies have linked gum disease to an increased risk of inflammation in the body and to memory impairment and dementia.[5] That may be because such conditions are often interlinked with other health problems–diabetes raises the risk of dental problems, for example–or brain-shrinking lifestyle habits.

Thyroid Conditions Hypothyroidism, a condition in which the thyroid gland doesn't produce enough of the hormone that controls metabolism, is very common among women in particular. But men can suffer from it too, and both can suffer from hyperthyroidism, in which the thyroid gland produces too much hormone. Hypothyroidism can contribute to a variety of problems, including obesity, which, as you'll soon read, is a brain shrinker. Hyperthyroidism, meanwhile, can cause nervousness, fatigue, or even heart failure, which also shrinks the brain.[6]

Cardiovascular Problems High blood pressure, high cholesterol, and diabetes can all lead to cardiovascular problems, which increase the risk of heart attack and stroke. In addition, poor blood flow to the brain dramatically shrinks the brain.[7] Although heredity plays a part in these risk factors, all are greatly reducible through lifestyle changes, and treating these conditions has been shown to reduce the risk of cognitive decline.

Liver Damage Alcohol abuse and certain other conditions can cause liver damage, which results in increased ammonia in the blood and indirectly damages the brain. In addition, alcohol abuse by itself, even without liver failure, directly causes brain atrophy or shrinkage, especially in the pathways linked to memory and attention.

Obesity Excess weight and obesity–particularly in the form of belly fat–are known brain shrinkers, reducing cognitive function in the short term and raising the risk of dementia in the long term. Responsible weight loss or healthy weight maintenance almost always involves dietary changes and exercise, both of which can grow the brain.

Menopause Menopause, the end of a woman's fertility, typically occurs between the ages of forty-five and fifty-five and can cause sleep problems and brain fog—memory lapses, slow thinking, or mild confusion—in some women. It's important to note that menopause has not been shown to directly shrink the brain and that its associated problems disappear when menopause does.

Sexual Dysfunction Patients often shy away from talking about this, but when I ask—and I always ask, because of its importance to overall health—I often uncover a problem. Sexual dysfunction can be a sign of another treatable health problem, such as heart disease, diabetes, or depression, or a side effect of medications. If left untreated, sexual dysfunction can affect your sense of well-being and joy. Although some people have intimacy without sexual interaction and are perfectly happy about it, many find sexual dysfunction—and the decrease in self-confidence it often brings—leaves them feeling frustrated and anxious, driving them away from the ideal alpha zone.

Urinary Frequency Kidney, prostate, or bladder issues can cause a host of problems, including urinary frequency, which might contribute to fragmented, poor sleep. (If you wake up multiple times each night to use the bathroom, you may have a bladder problem.) Long-term poor sleep can ultimately result in a loss of synapses on a microscopic level.

Allergies Food allergies such as gluten sensitivity may cause a host of symptoms that may impact the brain, from headaches to brain fog. Seasonal allergies can also cause headaches, congestion, interrupted sleep, and brain fog.

Back Pain Back pain can have many causes, and it can be exacerbated by stress or depression. Often people suffer through back pain and fail to seek treatment. However, chronic pain has been shown to damage brain pathways and shrink the brain.[8] Pain can also keep you from exercising, thus robbing you of a real brain builder. Fortunately, back pain is often treatable, and eliminating the pain can reverse the brain-related problems it causes.

Ankle Edema Edema, often showing up as swollen ankles, may be a sign of kidney or even heart problems, including heart failure, which has been shown to shrink the brain. Edema is treated by treating the underlying condition.

Vitamin D Deficiency People often forget that the skin is an organ, and one that's integral to our health. When it comes to brain health, the biggest concern related to skin is low vitamin D, a condition that can be caused by inadequate exposure to sunshine. Since overexposure to the sun isn't healthy either, I recommend knowing your vitamin D level and taking a vitamin D supplement (with calcium).

Numbness Reduced sensation or tingling in the toes can be one of the first symptoms of diabetes, which as you'll read in chapter 11, is a known brain shrinker.

Mental Health Conditions

A healthy body is critical for a bigger brain, but so too is a brain that's free of disease. There are a host of medical conditions related to the brain itself, but I'll limit this discussion to those most common and most important to improving cognitive function and building brain reserve.

Depression Depression is a bona fide brain shrinker that has been shown to reduce the size of the hippocampus and lead to a reduction in cognitive performance. Depressed patients often report memory problems, difficulty focusing and sustaining attention, and low motivation, among other symptoms. Fortunately, treatment of depression has been shown to reverse hippocampal shrinkage in just six months.[9]

Anxiety Anxiety is known to increase the risk of insomnia, stroke, and heart attack. It can increase levels of cortisol, which shrinks the hippocampus in particular. Meditation, stress reduction, cognitive behavioral therapy, and neurofeedback can all be helpful in reducing anxiety, as can medications.

Attention Deficit Disorder ADD is actually called attention deficit hyperactivity disorder-inattentive (ADHD-I), though it's still often unofficially referred to as ADD, especially in adults, who tend to lack the H of hyperactivity. ADD is characterized by inattention, distractibility, disorganization, and forgetfulness. Like those with depression and anxiety, people with ADD typically exhibit brain wave activity that's far from the norm—in this case showing too much slow theta activity in the frontal lobes, the area of the brain associated with paying attention and decision making. While it may seem minor, an inability to focus can wreak havoc on a person's cognitive performance, as many ADD sufferers can attest. Fortunately, treatment for ADD can be highly effective. Neurofeedback is one option that increasingly shows great promise for long-lasting treatment of the condition.

Body Scan

Complete the following body scan, checking off the items that apply to you and the conditions that you consider a priority. (Hint: These should be the

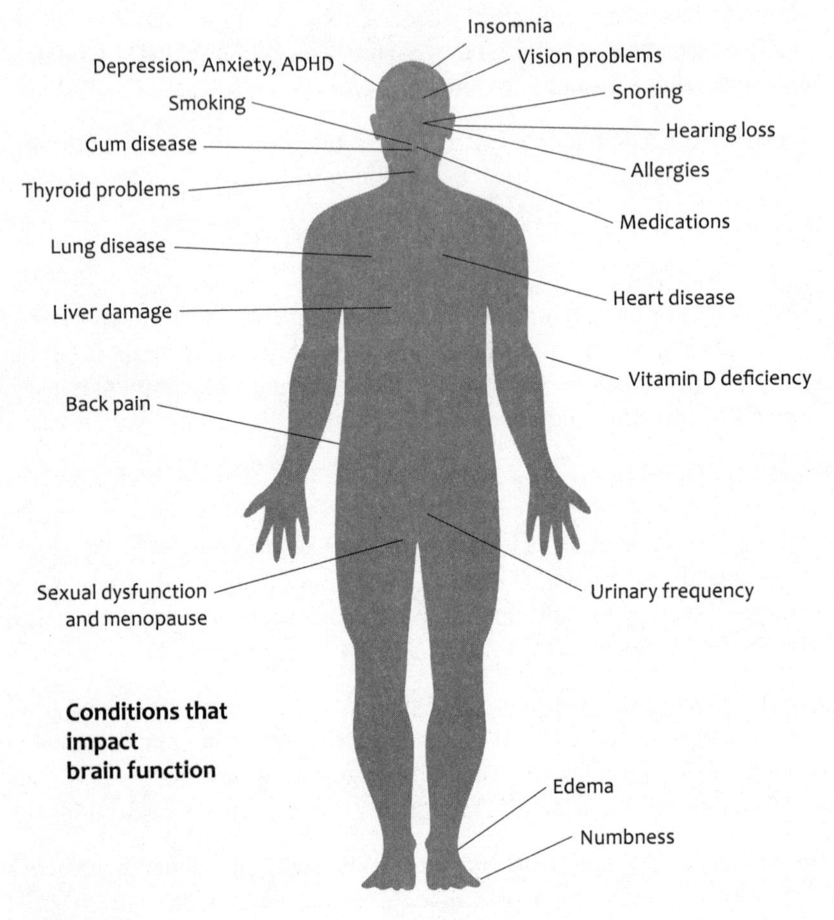

Body Scan

To check your brain's health, you first need to check your overall health from head to toe.

conditions that are most severe for you and most likely to take a toll on your brain's health.)

Once you've completed your body scan, make an appointment with your doctor to discuss the conditions you've identified as priorities. (You can talk about the others, too, if you'd like.) Be proactive: think about what you'll need to ask your doctor so that you can best address your problem areas. I recommend bringing a written list of four or five questions, with space to jot down your doctor's comments. You'll find space for questions in the area that follows.

Your Body Scan

Health conditions	✓ Applies to you	✓ Priority
Hearing loss		
Vision problems		
Insomnia		
Snoring		
Medication side effects		
Smoking		
Lung disease		
Gum disease		
Thyroid problems		
Cardiovascular disease		
Liver damage		
Obesity		
Menopause		
Sexual dysfunction		
Urinary frequency		
Allergies		
Back pain		
Ankle edema		
Vitamin D deficiency		
Numbness		

Mental health conditions	✓ Applies to you	✓ Priority
Depression		
Anxiety		
ADHD		

Questions for My Doctor

Here's an example of a condition and the questions you might have for your doctor. Below, fill in other conditions of concern to you.

Problem or Condition: Obesity

1. How bad is it?
2. How much weight should I lose?
3. How do I do it?

Problem or condition: ____________________

1. ____________________
2. ____________________
3. ____________________

Problem or condition: ____________________

1. ____________________
2. ____________________
3. ____________________

Problem or condition: ____________________

1. ____________________
2. ____________________
3. ____________________

Problem or condition: ____________________

1. ____________________
2. ____________________
3. ____________________

Problem or condition: ____________________

1. ____________________
2. ____________________
3. ____________________

Set Your Priorities

In the coming chapters, I'll give you details of how certain lifestyle changes can immensely increase the size of your brain. Their incredible brain-growing potential, of course, is tempered by the existence of negative health conditions we know shrink the brain.

I want to get you started thinking about changing your life. And—you already know this, I'm sure—it won't just happen by itself. You'll need to change priorities and squeeze time into your schedule for brain-boosting activities, while also committing to reducing your exposure to brain shrinkers.

Use the following chart, and check off the items you believe will require more of your attention. As you work through your twelve-week plan, refer back to this page as a reminder to maintain your focus on addressing these priority areas.

Brain growers	Needs my attention?	Brain shrinkers	Needs my attention?
Exercise		Excess weight or obesity	
Healthy diet		Smoking	
Meditation		Hypertension	
Memory training		Diabetes	
DHA		Head trauma	
Positive attitude		Stress	
Physical fitness		Alcohol abuse	
Learning new hobbies		Sleep disorders	

Meet the Brain Meter

As you read about brain growers and brain shrinkers in the pages ahead, you'll probably begin to wonder how much each affects the brain. Does exercise grow the brain more than meditation? Is sleep apnea worse for your brain than stress? How big of a boon is cognitive stimulation?

To help you gauge the relative value of each, I have developed a visual representation—the Brain Meter—which offers an indication of the potential impact each brain grower or shrinker has on the brain. The exercise Brain Meter, for example, indicates the benefit of average intervention and the maximal

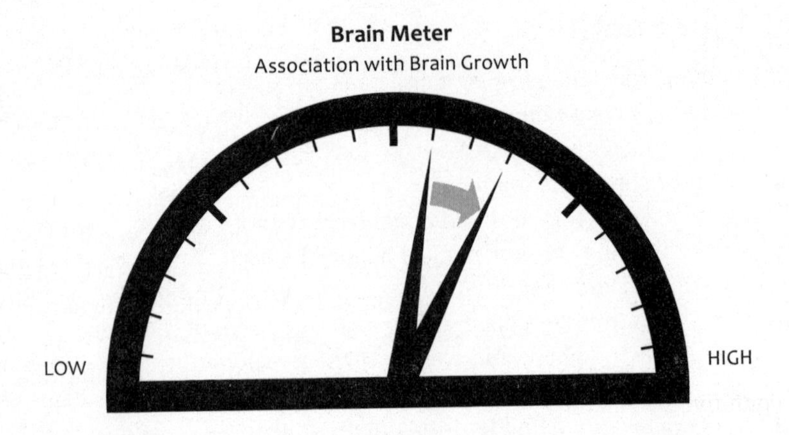

brain benefit of exercise. You'll find Brain Meters in each chapter that deals with a brain grower or brain shrinker. Use them as a guide to help you tailor your brain fitness efforts.

Your Rx: Choose Your Track

It's easy to get overwhelmed by a task as big as changing your life (or growing your brain!). Where do you start? What do you do and when do you do it? In chapter 8, I'll give you my twelve-week prescription for a bigger, better brain. But, as I detail the benefits of brain growers in part II, I'll give you a preview of the precise instructions you'll use to maximize each brain booster in your own life.

Of course, since there's no one-size-fits-all brain, it stands to reason that there's no single approach to building brain reserve. A thirtysomething who loads up on fruits and vegetables, with a healthy smattering of fish thrown in, won't find it very helpful if he's told to eat less fast food. Instead, his high-yield brain fitness approach might revolve around adding supplements to his diet and targeting certain fruits and vegetables in particular that are known to boost oxygen flow, BDNF, and even promote healthy brain activity.

His less healthy peer—say, a fifty-year-old who eats a high-fat dessert four times a week, regularly downs potato chips, and takes two heaping servings of red meat at every meal—requires an approach that looks quite different. For him, the first priority will be cutting out sweets and junk food, limiting portion sizes, and ensuring half his dinner plate is devoted to fruits and vegetables. As the weeks progress, he can focus on a diet with maximum brain-boosting effects. Similarly, while I might recommend forty-five minutes a day of aerobic activity for a fairly fit person starting the program, I'd start his sedentary friend with ten minutes of walking, three times a week.

You might find yourself in need of improvement in every aspect of brain fitness. Or in just a few. To help you customize your plan, I've designed a track system that will allow you to select a starting pace in each brain booster category. As you complete each chapter in part II of this book, you'll be asked to select your starting track in that area (track 1 for beginners, track 2 for intermediates, track 3 for experts). You'll likely find yourself on different tracks for different brain boosters. If you do mental gymnastics daily but rarely work up a sweat, for example, you'll likely select track 1 for exercise and track 3 for cognitive stimulation.

In general, the tracks are divided as follows:

Track 1

If you're far from fit in a particular brain-boosting area (for example, you scored a 1 in the relevant areas on the brain fitness calculator), you'll be on track 1. In this track, you'll start slowly and work your way up to a target pace over the first three weeks of the course. After three weeks, you'll move up to track 2.

Track 2

If you're moderately fit in a particular brain-boosting area, you'll be on this track, which will start at a slightly more advanced level than track 1. After week three, you'll move up to track 3, or repeat the third week of track 2 until you're ready to advance.

Track 3

If you're already fairly well positioned in a particular brain-boosting area, track 3 will offer you high-yield tips on how to fine-tune your brain fitness. Once you reach week three you'll continue with the brain-boosting recommendations of that week for the remainder of the program, adding intensity on your own based on what I've taught you.

Look back at your brain fitness score and choose a beginning track for yourself in each of the following booster areas. (You'll refer back to this page as you progress through your twelve-week plan.) Then choose a goal track. This is the track you'd *like* to be on as you reach week twelve. Be realistic. While I'd love to see everyone on track 3 by the end of three months, I recognize that those who have further to go, or who have health constraints, might take much longer to get there. What's important is that you're developing habits that will continually propel you toward a bigger, better brain. You can reference this page when you reach chapter 8 and start your twelve-week plan.

Choose your tracks:

Brain Booster

Date:	Beginning track (choose 1, 2, or 3)	3-month goal track (choose 1, 2, or 3)	1-year goal track (and beyond)
Exercise			
Diet			
Mindfulness			
Cognitive stimulation			

PART II

Brain Boosters

CHAPTER FOUR

The Fit-Brain Workout

MY COAUTHOR, Christina, has a morning routine that unfolds like clockwork. At six A.M. each day she is up and out of bed, ready to tiptoe quietly down a flight of steps to her kitchen in the hopes of sneaking in a few solitary minutes before the kids awaken. If all goes well, she'll read the newspaper, eat a healthy breakfast, and then ready herself for the onslaught of energetic little ones.

The sequence has been the same for years, except that in the winter of 2012, Christina altered it in one critical way. She still had the morning paper on her mind as she rolled out of bed, but before she headed for the stairs she added a new step: grabbing a small electronic device from her bedside table and clipping it onto the waistline of her pajamas. As she ambled toward the kitchen, the device—no larger than her thumb—recorded the forty-two steps she took to get there. It then logged the sixty steps she took puttering around downstairs before climbing the stairs (which the device would also record) to wake her kids for school.

The gadget—this one is called Fitbit, but there are several on the market—uses motion detection software, similar to that used in Wii games, to calculate Christina's steps as well as count how many flights of stairs she conquers each day. It also wirelessly transmits the data to her computer and logs it in a software program she can use to help track her activity. (The software also allows her to enter everything she eats or drinks and provides her with a running count of the number of calories she's consumed and how many she's burned.)

Forty-three-year-old Christina, a mom of three, was hooked the moment she first clipped on her high-tech pedometer. The tiny step counter, she found, offered the encouragement she desperately needed after eight years in which her primary form of activity was ushering her kids in and out of the car for the drive to soccer or school or kung fu. Suddenly, every step seemed to count. Weighing in at the high end of normal BMI, Christina hoped adding extra steps to her day would help her shed a few pounds. If upping her activity would also help sharpen her thinking and put her on a path to future good health, it would be a bonus well worth the effort.

Christina's eight thousand or so steps a day began to creep ever upward as she made a greater effort than ever to exercise. Within two weeks of carefully monitoring her activity, she upped the intensity, signing up for a half marathon as an extra motivator. With that, she had three months to train—just the right amount of time to implement a brain fitness program.

Running several times a week would surely help her meet her weight goal. But the question remained: Could moving her body really make a difference in the way her mind moved? There was one way to find out: Christina underwent cognitive testing and an MRI before starting my brain fitness program and then repeated the tests at the close of the three months.

I knew what to expect; even so, I was surprised by the degree of change I saw. In the course of the three months, Christina's hippocampus had grown a whopping 5 percent. Keep in mind that she was still relatively young and relatively healthy, even before she started the program. Her preprogram lifestyle didn't expose her excessively to brain shrinkers. To see such dramatic growth in her hippocampus, then, was truly amazing.

And what about her cognitive function? The results of Christina's paper-and-pencil tests showed more good news: her short-term memory had jumped 15 percent.

It's nearly impossible to tease out how much of Christina's improvement was attributed to exercise and how much to other factors, like the DHA supplement she began taking, the regular breathing exercises she added to her day, the ten pounds she lost during the study period, or the intensive cognitive stimulation of helping me write this book. But I believe exercise had a significant impact. Her larger hippocampal size was, most likely, a result of a boost in her BDNF, a spike in the amount of oxygen and in the number of new blood vessels in her brain, and the healthy brain activity she promoted by changing her lifestyle. As you'll soon read, there's a great deal of recent research on which to base these assumptions.

Healthy Hippocampus

If you remember my discussion of neurogenesis from chapter 2, you'll recall the research by Fred Gage and others that showed neurogenesis in action in the brains of first animals and then humans. Gage's research on mice in enriched environments was exciting, but even more exciting was what happened in 1999, when Gage and his team peeled back another layer of the mystery of neurogenesis.

This time, the research team teased out the effects of one particular intervention in the lives of those enriched mice: exercise. Remember, mice who lived in cages with social interaction, toys, and running wheels saw more of their new neurons survive. Exercise alone, however, dramatically increased the number of new neurons being born.[1] Looking at the slides under a microscope, "we were really kind of struck," Gage recalled for my co-author, Christina, in an interview. "It was actually surprising that running alone could have such a robust effect. You could hold the slide up to the light and see the difference in the brains."

The next question was, of course, what effect did exercise have on the brains of humans? It's an area of research that had already been broached from another angle in observational studies. Researchers had long noted that exercise reduced the risk of dementia. But only in recent years have we begun to understand what, exactly, is going on in the brains of those who exercise.

Some of that insight comes from University of Pittsburgh Assistant Professor Kirk Erickson. I often see him at medical conferences and we chat about the incredible malleability of the human brain and how easily it grows. He has studied the topic in depth, and has repeatedly tied exercise to a larger hippocampus and better cognitive performance. In one study, published in 2009, Erickson and his colleagues documented larger hippocampi in older adults who had higher fitness levels.[2] Not only did MRIs reveal the structural changes, but cognitive testing showed those higher volumes also had a real payoff: people who'd exercised more performed better on tests of spatial memory skills.

But was it really exercise that caused the hippocampus to grow? Erickson and his team aimed to find out. To do so, they enlisted 120 cognitively healthy elderly adults aged sixty to eighty who weren't physically active at the start of the study.[3]

Researchers split the participants into two groups, with one group assigned to a regular routine of brisk walking—forty minutes, three times a week—while the other group did stretching and toning exercises. All were supervised by trained personnel, who monitored their heart rates and levels of exertion. And both groups underwent MRIs that measured the size of their hippocampi and blood tests that measured their BDNF levels.

A year later, study participants underwent another round of MRIs to see if their brains showed the effects of their efforts. Those who'd engaged in aerobic exercise saw their hippocampi grow by about 2 percent over the year they'd been exercising. Considering that the hippocampus normally shrinks by 0.5 percent per year as we age, study participants had effectively walked away as much as four years of brain aging. And those who showed greater changes in their BDNF levels saw greater increases in their hippocampal volume—a sign that it was BDNF that was spurring growth in the hippocampus.

By comparison, study participants in the stretching group—who on average didn't improve their overall fitness—experienced about a 1.4 percent *shrinkage* in their hippocampi over the study period, about what you'd expect for this age group. Those who started out more fit, however, saw less decline—yet another argument for building brain reserve and keeping your hippocampus plump.

How did these changes affect participants' cognitive performance? It was no surprise to find that the walkers did better in memory-function tests than their stretching peers.

Instead of merely noting a tie between fitness and hippocampal size as his earlier research had, this time Erickson had proof that "an exercise intervention can actually increase the size of brain regions—especially the hippocampus," he says. In other words, you *can* take an inactive elderly person, make him or her active, and watch that person's hippocampus grow.

Erickson, in fact, calls it "one of the really important messages" from the study. "Starting an exercise regimen in late life after being physically inactive for most of your life is not futile!" he says. "You can still reap some of the benefits from exercising."

Most incredible of all, "we know of no medication that can do that," says Erickson. I agree; this is incredible indeed.

Of course, as reassuring as it is to know that exercise has benefits even in late life, we know, too, that it has clear brain-growing capabilities throughout life.

Children who are fit, for example, have larger hippocampi than their less fit peers, as Erickson and his colleagues demonstrated in a study whose results were published in 2010.[4] For that study, the research team recruited forty-nine nine- and ten-year-olds and measured their oxygen consumption (also known as VO_2 max)—a commonly used measurement of physical fitness—while they ran on a treadmill. All the children then underwent MRI scans.

For Erickson, who'd by then seen the effect of exercise on the brains of the elderly, the results weren't a given. In late life, when we know the hippocampus is shrinking, "it makes more sense to find an effect where fitness would reduce atrophy," Erickson explains. "I actually was wondering if we were going to see the same effect in kids, just because kids' brains are developing at this time."

But even in young brains—which are growing rather than shrinking—exercise can clearly affect the *rate* of growth. In the children studied, those who were more fit had hippocampi that were 12 percent larger relative to total brain size than their less fit peers.

"The fact that we were able to detect this fitness effect just goes to show you the robustness and the size of fitness and physical activity on the developing brain," says Erickson. "It's quite impressive." Fitter children also did better on memory tests than their less fit peers.

There's now compelling evidence that exercise has a direct role in increasing BDNF levels too. In 2011, for example, researchers at the University of Dublin linked increased exercise with higher levels of BDNF—and better cognitive performance.[5] For that study, the research team recruited sedentary college students and gave them a face-name matching test. Then, half the students climbed aboard stationary bikes and engaged in a bout of strenuous cycling, while the other half rested. Both groups then took the memory test again and were assessed for levels of BDNF through a blood test.

Participants who engaged in a short period of high-intensity cycling did better on the face-naming task after exercise than they had before it. Those who hadn't exercised did about the same on the test as they had the first time. The group that had cycled also had higher levels of BDNF in their blood than they'd had before exercising, while the non-cycling group did not.

Five weeks later, with the cycling group continuing to cycle regularly, the team performed the tests again. They found that five weeks of cycling training increased BDNF levels and performance on the face-naming test, leading researchers to conclude that both acute *and* repeated exercise benefit the hippocampus.

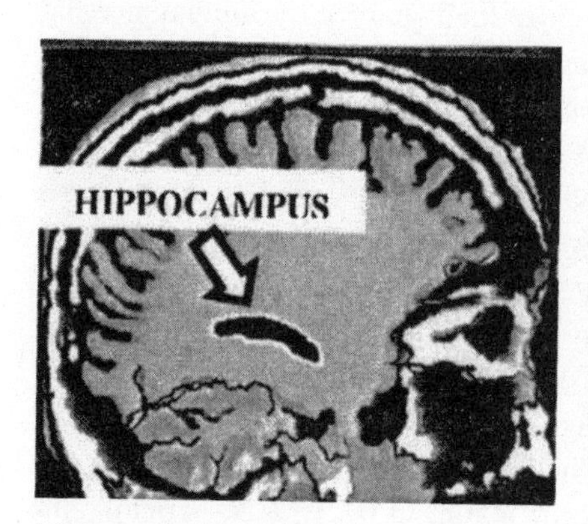

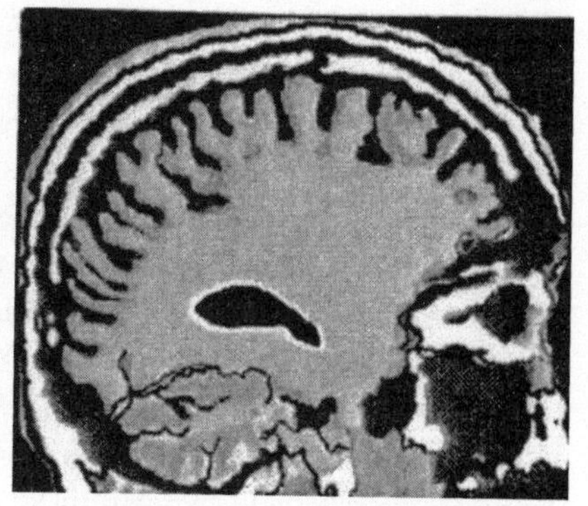

Hippocampal Growth

Before and after images illustrate the type of growth that occurs in the hippocampus with vigorous exercise.

And the Cortex

Having a larger hippocampus, of course, is only part of the story. There's plenty of evidence that exercise also increases the size of the cortex. One such study comes from Erickson and his colleagues, who examined 299 non-demented elderly people enrolled in the Cardiovascular Health Cognition Study, a longitudinal study that gathered data from 1988 to 2009.[6]

Grouping the participants by the number of blocks they reported walking each week, the team studied participants' initial physical assessments and then followed up with MRI scans nine years later. Four years after that, the team tested the participants for cognitive impairment and dementia.

The results were impressive. Those who'd walked regularly—about six to nine miles a week—had significantly more grey matter in the frontal, occipital, and hippocampal regions than those who walked less. Checking in thirteen years after participants' initial assessments, Erickson's team found that those who'd logged six to nine miles a week were far less likely to be cognitively impaired than those who walked the least.

Other research has shown similar changes in grey matter as well as in white matter associated with exercise. In one randomized clinical trial of fifty-nine cognitively healthy people between sixty and seventy-nine years old, for example, researchers at the University of Illinois found that study subjects who engaged in six months of aerobic exercise had more grey and white matter in the frontal and temporal lobes than when they'd started the study. Those who merely did toning and stretching exercises saw no such increase.[7]

In 2012, University of California, Los Angeles (UCLA), researchers offered yet more evidence, presenting the findings of their study at the annual meeting of the Radiological Society of North America. Led by Dr. Cyrus Raji, the team examined the lifestyle habits and MRIs of 876 adults aged sixty-nine to ninety-five. Study subjects had varying degrees of cognitive health—some had normal cognitive function, some had mild impairment, and others had Alzheimer's disease. The team calculated how many calories study subjects burned doing physical activities that ranged from sports, to yard work, to dancing, to cycling. Activity levels varied dramatically: those in the top twenty-fifth percentile burned about 3,400 calories a week, while those in the bottom twenty-fifth percentile burned just 384 calories a week on physical activity.[8]

What was going on in the brain was also dramatic: those who'd burned the most calories had 5 percent more grey matter in the frontal, parietal, and temporal lobes—including the hippocampus—than those who burned the least.

A Combined Effect

We're still working to understand exactly how exercise grows the brain. Increased BDNF no doubt is a major part of the story, especially when it comes to growth in the hippocampus. But there are other reasons exercise grows the brain.

For starters, exercise promotes cardiovascular health, helping the heart pump blood more efficiently to all parts of the body, the brain included. One way it does this is by increasing levels of high-density lipoproteins (HDLs), the "good cholesterol" in the blood, which allows blood to flow more freely. Another, as you'll recall from the first pages of this book, is via a process called angiogenesis, the creation of new blood vessel branches. Exercise promotes angiogenesis. That means that as you exercise you're actually adding to the brain's network of blood vessels, bringing more oxygen to every part of the brain.

In fact, if you look at the network of blood vessels in the brains of exercising adults, as one 2009 research study did, you'll find it far more extensive than the blood vessel networks that exist in the brains of their sedentary peers.[9] For that study, the research team performed magnetic resonance angiography (MRA) on the brains of fourteen healthy elderly participants. Seven of the participants reported a high level of aerobic activity—at least three hours a week over the prior ten years—while the other seven reported exercising less than ninety minutes a week over the prior ten years.

Looking at the results, the study team found that those who had exercised more had healthier-appearing blood vessels and more branchings of their small blood vessels. The difference between the two sets of brains is striking—and the effect might even have been greater than we can see on MRA, since some of the brain's blood vessels (called capillaries) are so tiny that they can only be measured through a microscope.

A healthy vascular network with dense branches of blood vessels means more nourishment for neurons, helping them to stay vibrant and alive, and better retention of the brain's highways, allowing different parts of the brain to communicate more efficiently. The newly formed and engorged blood vessels may also account for part of the remarkable growth in the size of the hippocampus with exercise.

In addition, exercise aids the brain in another way: by promoting the creation of new mitochondria, the powerhouses of energy within cells. These new mitochondria are born throughout the body but are put to use in different ways. In muscles, for example, they power the processes needed to add muscle mass, bulking up your biceps when you lift weights or your calves when you run.

In the hippocampus (which I sometimes think of as our "memory muscle"), mitochondria help fuel the growth of new cells, new synapses, and new small blood vessel branchings. The additional energy they provide is critical to rejuvenation and repair elsewhere in the brain, too. I believe mitochondria are the ultimate brain revitalizer—the more we have, the longer we'll live and the sharper we'll be in old age.

Running to Alpha

There's one other factor in how exercise changes the brain: healthy brain activity. Exercise can promote healthy brain wave activity by increasing activity in the healthy alpha range, which is associated with a calm, alert, focused state of mind.

In one study, researchers in Germany and Australia used EEG to measure the baseline brain wave activity of twenty-two recreational runners between the ages of twenty-one and forty-five.[10] The study participants then ran on a treadmill. When they finished, their brain wave activity was again measured via EEG. What the research team found: in all the study subjects, alpha brain wave activity increased immediately after exercise.

Alpha activity was especially strong in the left frontal lobe—the part of the brain that's most closely associated with focus, attention, and decision making. That this occurs will be no surprise to anyone who has ever experienced the rush of a "runner's high." Part of that feel-good effect is due to the release of endorphins and dopamine, neurotransmitters that promote a sense of well-being. It's possible that increased levels of these or other feel-good brain-messenger molecules promote alpha activity in the brain. Or that the rush of oxygen-rich blood flooding the brain during exercise is what's behind that increased alpha activity. Or, more likely, a combination of factors.

I am looking forward to seeing additional solid evidence detailing the link between exercise and healthy brain activity; in the meantime, it's clear that exercise does indeed help us enter the alpha zone. For proof, all you need to do is consider how *you* feel after vigorous exercise.

Faster? Stronger?

There's still much research needed to determine how much exercise—and of what type—is most beneficial. For now, however, it's clear that vigorous aerobic activity offers brain benefits. Walking is better than not walking. But adding intensity—and resistance—may offer a robust advantage.

One recent small study presented at the Canadian Cardiovascular Congress offered evidence that a program of high-intensity interval training, plus resistance training, may help improve cognitive function.[11]

For the study, six middle-age men and women participated in a four-month exercise program, which included two days a week of high-intensity interval training (HIIT). The training consisted of thirty minutes of exercise followed by ten minutes of high-intensity interval training, which involved intervals of thirty seconds of sprints and thirty seconds of lower-intensity cycling. Participants also had twice-weekly resistance training sessions.

After four months, the HIIT participants showed a 10 to 25 percent improvement in cognitive processing speed and short-term memory, as well as increased blood flow to their frontal lobes. This may be an indicator that short bursts of intensity promote mitochondrial biogenesis in the brain, just as they do in muscles.

One shortcoming of the study was that it did not include a control group that had only continuous moderate exercise, without HIIT. However, other studies have shown high-intensity exercise significantly increases the number of mitochondria,[12] so I favor adding short bursts of HIIT at the end of my own fitness training.

The Stroke Link

You'll read in chapter 12 about the devastating effects of a stroke, which is, without question, one of the biggest brain shrinkers. Imagine if we had an intervention that proved highly effective in reducing the risk of stroke. What if it were free? What if it were accessible to almost everyone, at any age?

You've guessed where I'm going with this, I'm sure. We do have such an intervention: exercise. Physical activity can go a long way to reducing a variety of vascular risk factors, which in turn reduce the risk of stroke. And it's not a small effect.

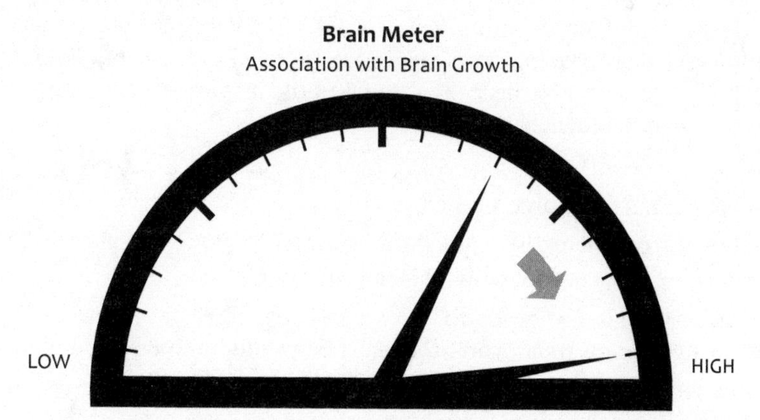

Exercise boosts BDNF, enhances oxygen flow, and promotes healthy brain activity. The more you increase your fitness, the more you will build a bigger "memory muscle" in your hippocampus. For optimal brain growth, I recommend thirty minutes of vigorous aerobic activity, plus fifteen minutes of resistance training, five days a week.

In one study published in 2011, a team of researchers looked at 1,238 stroke-free participants enrolled in the Northern Manhattan Study.[13] Participants' average age was seventy, and all underwent MRIs to check for silent strokes. They were then sorted into quartiles based on their physical activity levels. Compared to those who didn't exercise, those in the highest quartile exercise group were almost half as likely to show small strokes on their MRIs.

Your Rx

You need to exercise, but don't head to the gym until you have the "all clear" from your doctor. In the meantime, here's a preview of the tracks you'll choose from in chapter 8 as part of your brain fitness program:

Track 1

If you lead a sedentary lifestyle and are currently not exercising at all, this is your track. I want you to exercise, but I don't expect you to run a marathon (at least not anytime soon). As excited as you are, be sure to start slowly. You don't want to risk an injury. If you have physical limitations, talk to your doctor about substitutions for walking or jogging.

Week One: Start by walking for ten minutes at a time, three days a week.

Week Two: Increase your walking time to fifteen minutes, three days a week.

Week Three: Increase your walking time to twenty minutes, three days a week. In addition to your scheduled time, find ways to work activity into every day. Take the stairs, park at the far end of the parking lot, walk to lunch, go for a hike instead of going to the movies—do anything you can to replace sedentary moments with activity.

Track 2

If you're not completely sedentary—for example, if you walk a good bit for work—and exercise occasionally, this is your track.

Week One: Walk fast or jog for twenty minutes, three days a week. (Don't count time you walk at work; this must be twenty uninterrupted minutes.)

Week Two: Increase your walking or jogging time to thirty minutes, four days a week. If you're walking, try to jog at least some of the time. Increase the amount of time you jog versus walk as the weeks progress.

Week Three: Increase your walking or jogging time to thirty minutes, five days a week. Gradually increase the intensity of your workouts. Ideally, you want to get to a level of intensity to achieve 60 to 80 percent of your maximum heart rate. If you'd like, you can substitute a stationary bike, swimming, or another aerobic activity, or you can substitute a one-hour game of tennis (or another intense activity) for two of your thirty-minute sessions.

Track 3

If you already exercise twice a week or more, this is your track.

Week One: Do thirty minutes of uninterrupted aerobic activity, five days a week, plus five extra minutes of weight lifting, push-ups, or another muscle-building activity, three days a week.

Week Two: Continue with thirty minutes of uninterrupted aerobic activity, five days a week. Bump up your muscle building to ten minutes, three days a week.

Week Three: Continue with thirty minutes of uninterrupted aerobic activity, five days a week. Bump up your weight lifting to fifteen minutes, five days a week. Increase the intensity of your workouts as needed so that you're improving your physical fitness. Consider adding high-intensity interval training (HIIT), consisting of five bursts of vigorous exercise during the last fifteen minutes of your exercise routine (five bursts for one minute each, followed by two minutes of less intense activity). You should feel tired after you finish your exercise.

CHAPTER FIVE

Your Recipe for a Bigger Brain

IF I WERE TO TELL you that what you eat affects your brain health, you would probably roll your eyes. We've all heard—for years!—about brain food. We know we're supposed to eat blueberries and spinach. We know that they contain antioxidants and will counteract inflammation in our bodies. We know that in some mysterious way this is supposed to help us stay mentally sharp as we age.

What most people *don't* know is that the food you eat literally reshapes your brain.

Scientists have long had an inkling of this. Way back in 1972 Romanian psychologist and chemist Corneliu E. Giurgea used the term "nootropic" to describe substances that stimulate nerve growth or affect the brain's neurochemicals or oxygen supply. Since then, a growing body of research has added solid proof to the notion that what we eat affects our brains' structure and size—and therefore how, and how well, we think.

And while inflammation undeniably has an impact on our long-term brain health, it's nowhere near the only factor at play. Just as exercise does, brain foods work their magic by boosting BDNF and increasing blood flow to the brain. Some, like green tea, even heighten healthy brain activity.

The end result is a thicker cortex and a larger hippocampus, which together add up to a more fully "funded" brain reserve that serves us both now and in the future.

But growing your brain through diet isn't just a matter of adding a few key ingredients to your food or popping a few pills. A brain-boosting diet also steers you clear of known brain shrinkers, such as obesity, diabetes, high blood pressure, high cholesterol, metabolic syndrome, and stroke (you'll read about these in part IV).

In fact, the best recipe for a bigger brain starts with a healthy, balanced diet and then adds to it a serving of the key nutrients—DHA, plus select flavonoids and vitamins—we now know can grow the brain.

Your Brain . . . on Food

Let's start with that basic healthy diet. Many studies have shown that the Mediterranean diet—a diet low in fat and cholesterol and high in fiber—omega-3 oils, vegetables, nuts, and fruit, can help stave off cognitive decline. People who adhere to such a diet have a much lower prevalence of age-related cognitive impairment than those who don't.[1]

In one study, led by Columbia University neurology professor Nikolaos Scarmeas, this diet offered a roughly 28 percent lower risk of cognitive impairment and dementia.[2]

The Mediterranean diet also helped the already impaired elderly lower their risk of developing Alzheimer's disease, at least during Scarmeas's study period. Study subjects who had cognitive impairment and were in the top third for adherence to a Mediterranean diet had a 48 percent lower risk of progressing to Alzheimer's disease, as compared to people with cognitive impairment who were in the lowest third of adherence.

When combined with exercise, the effect can be even more powerful. When Scarmeas and his team followed 1,880 elderly New Yorkers over a period of fourteen years, they found that those who exercised regularly *and* adhered to a Mediterranean diet had a whopping 65 percent lower relative risk of developing Alzheimer's disease.[3] Of course, there may be other factors common to those who stay active and eat such a diet. Such people may be more inclined to engage their brains, or stay socially active and less stressed, or they are different in some other way that brings with it a neuroprotective benefit.

However, recent studies show certain foods do affect brain size, independent of other factors. Some of the more compelling research on this topic comes from Gene Bowman, an assistant professor of neurology at the Oregon Health and Science University, who studied the link between blood serum levels of specific vitamins and brain size and function.[4]

When I met Dr. Bowman during the 2012 Alzheimer's Association meeting, his passion for the subject was evident. Bowman had just published the findings of his study in the journal *Neurology* and was eager to share the details. For the study, Bowman reported on 104 men and women in their eighties who were enrolled in the Oregon Brain Aging Study, which has tracked participants since 1989. Looking at their blood samples, MRIs, and cognitive tests, Bowman found that people with high blood levels of vitamins B, C, D, and E had larger brains and performed better on memory tests than those with lower levels of those vitamins. Those with higher levels of omega-3s in their blood also were more likely to have greater brain volume, and they scored better on tests of executive function.

Cut the Salt!

Our bodies require salt to survive, but the typical Western diet contains far more salt than we need. In fact, while the U.S. Centers for Disease Control and Prevention recommends a daily intake of 1,500 milligrams, with 2,300 milligrams being the maximum recommended upper limit, the national average salt consumption is more than 3,400 milligrams per day.

Processed and packaged foods tend to be especially high in salt. Consuming too much salt can raise your risk of high blood pressure, heart disease, and stroke, all of which can shrink the brain.

Drop the Donut

Perhaps most telling of all was what Bowman found in the blood of those who had the lowest brain volume and scored poorly on tests of memory, processing speed, attention, and language skills. Not only did those study subjects have lower levels of the key vitamins and omega-3s but they also had higher levels of trans fats in their blood. Trans fats, you probably know, are unsaturated fats that can occur naturally but are most often obtained through processed foods—think fast food, donuts, and mass-produced cookies and cakes—that contain partially hydrogenated vegetable oil. Trans fats raise the risk of cardiovascular disease by increasing LDL (the bad cholesterol) and decreasing HDL (the good cholesterol). And we now know, thanks to Dr. Bowman, that trans fats are linked with a brain that is shrinking rather than growing.

Of course, it's not just trans fats that shrink the brain. Animal studies also show us the dangers of too much sugar in the diet. In one study, published in 2012 by researchers at UCLA, rats fed a steady diet high in fructose—and low in omega-3s—showed signs of developing insulin resistance and had the slowed cognitive performance that goes with it.[5]

For the study, researchers trained two groups of rats on a maze twice a day for five days. They then fed one group regular rat food plus a fructose-water solution for six weeks. A second group was fed regular rat food and a fructose-water solution plus omega-3 fatty acids in the form of DHA and flaxseed oil. After six weeks on the diet, the rats tried the maze again. The rats in the omega-3 group fared much better than their non-omega-3 peers, remembering the maze fairly well. The rats in the fructose group, meanwhile, struggled to find their way through the maze. In addition, their blood tests showed that they had developed resistance to insulin, a hormone that's crucial to glucose regulation and brain cell function.

A diet high in saturated fat and sugar has also been shown to reduce hippocampal BDNF levels in animal studies.[6] In addition, such a diet contributes to other conditions, such as diabetes, high cholesterol, high blood pressure, and stroke, which are known to shrink the brain. A poor diet also contributes to a greater risk of obesity, which, as you'll read in chapter 11, also inhibits the body's production of BDNF.

Even in the short term, however, poor food choices can affect how you think. The blood-sugar rush you get from devouring a brownie, for example, may be followed by a crash, leaving you low on energy and thinking sluggishly.

That's all the more reason to stick to a low-carb, high-protein, balanced diet and steer clear of brain-shrinking foods. I recommend shunning foods that are high on the glycemic index (for a list, visit the American Diabetes Association at www.diabetes.org), which will help you avoid the sugar spikes that mimic diabetes. In addition, I strongly encourage adding brain-building nutrients to your diet to help grow your brain. The end result will be increased energy and cognitive performance today and a bigger brain for the future.

So, what are those brain-building nutrients? While there's much debate—and much research under way—I've selected the best of them to discuss next.

Omega-3s: Adding DHA to Your Diet

The nondescript commercial strip twenty miles south of Baltimore doesn't look like much. Outside there's a simple sign that reads "DSM," a parking lot, some tidy shrubbery. It is tranquil, even a bit dull. Looks, of course, can be deceiving. Inside, there's plenty going on.

The first clue is an assortment of consumer products just inside the front door: yogurt, baby formula, milk, orange juice, bread, crackers—the list goes on. They come in different shapes and sizes, and from different manufacturers, but they all have one thing in common: each product has been supplemented with docosahexaenoic acid, a long-chain polyunsaturated omega-3 fatty acid better known as DHA.

In this case, it's not just any DHA but a type of DHA cultivated from algae in a process the scientists here developed in the late 1980s. Those scientists were all members of a team working for a Washington-area defense contractor and assigned to study potential uses of algae in long-term space flight for NASA. As they probed for possible applications of algae as a food source, the team began to identify the distinct health benefits DHA offered. DHA, they began to prove, had clear rewards for the heart, the eyes, and the brain. Before

Forget the Pyramid

In 2011, the U.S. Department of Agriculture said good-bye to the long-familiar food pyramid, replacing it with a new dietary guideline called My Plate (www.choosemyplate.gov).

My Plate uses a dinner plate graphic to give you a visual idea of what a balanced diet looks like—fruit and vegetables make up half the plate, proteins and starch share the other half and are complemented by a small portion of dairy.

long they had their own company, called Martek Biosciences Corporation (now part of DSM), and were well on their way to ensuring DHA would make its way into consumer products across the globe.

A believer in the benefits of DHA, I have served as a consultant to Martek and watched as the company developed its products and spread the word about the benefits of DHA. Their research facility in Columbia, Maryland, has carefully compiled evidence, and researchers worldwide have added their own proof, that DHA delays cognitive decline and improves cognitive health in both children and adults.

By the mid-2000s, DHA's role as a key health product was solidly established. In health food stores and even in regular grocery stores, sales of fish oil, omega-3s, and other fatty acids shot up as people got the message that better cognition might be within their grasp. By 2011, Martek—which launched in 1985 with just a handful of employees—had been snapped up by Netherlands-based DSM for a whopping $1.08 billion.

Martek's success was part of a huge rise in the supplement market in general. But behind the company's spectacular growth is increasing scientific evidence that its marquee product offers very real cognitive benefits throughout life.

A Fish Oil That Stands Above the Rest

DHA is just one type of omega-3 fatty acid. And while omega-3s in general have been shown to reduce the risk of heart disease and stroke—two benefits that have their own brain health implications—DHA is the only one that comes with robust proof that it's a brain builder.

Found in certain fish, algae, and supplements, DHA is a vital component of brain health in adults and brain development in children. DHA is also crucial for eye and heart health, and has shown promise as a protection against depression, cancer, and diabetes.

Like many other brain boosters, DHA helps to reduce inflammation in the brain and may reduce the aggregation of amyloid plaques associated with Alzheimer's disease, although we can't yet say that it helps to prevent the disease. Researchers have also tied low DHA levels to serious conditions and problems—lower-than-typical levels are found in adolescents who are incarcerated and have behavior problems, in children with ADHD, and in people with dementia and Alzheimer's disease.

On the flip side, we have ample evidence that its presence is beneficial—DHA, along with another omega-3 called EPA (eicosapentaenoic acid), has a proven effect on heart health, increasing oxygen flow throughout the body, including the brain,[7] and has also been shown to increase BDNF and neurogenesis in the hippocampus.

In short, DHA helps to grow the brain. The result, of course, is improved memory and enhanced learning.

The evidence is laid out in a host of studies. One, published in 2009, was the type of randomized, controlled interventional study that scientists love best. It showed the effects of DHA on ninety-one South African schoolchildren, who for six months ate a bread spread enriched with 127 milligrams of DHA. By the end of the study, the children showed higher blood levels of DHA and significant improvement on cognitive tests that measured short-term memory and processing speed.[8] The ninety-two children in a control group, who did not eat the bread spread, showed no such improvement.

In another study, published in 2010, researchers in Gothenburg, Sweden, tracked the academic achievements of 9,448 schoolchildren and their dietary intake of fish.[9] What did they find? Children who ate more than one serving of fish per week had an overall school grade 15 percent higher than those who consumed less than one serving per week. Even eating just one serving of fish a week proved beneficial. Those students (about 56 percent of the total group) also scored higher than students who ate less than one serving a week.

University of Pittsburgh researchers found similar results in their study of 280 adults between the ages of thirty-five and fifty-four.[10] That study's participants underwent cognitive testing and had blood samples drawn to check omega-3 levels. Those with higher DHA levels scored better on nonverbal reasoning and mental flexibility tests as well as on working memory and vocabulary measurements.

One of the best studies to show the benefits of DHA was MIDAS (Memory Improvement with Docosahexaenoic Acid Study), conducted at nineteen

DHA: Your Daily Dose

While I always favor real food over supplementation, the truth when it comes to DHA is that it's almost impossible to get as much as you need through diet alone. Therefore, I recommend eating two servings of fish per week (fatty fish like salmon, halibut, and tuna are the best sources of DHA) and supplementing with a daily dose of 1,000 milligrams of DHA (100 to 200 milligrams for children).

When buying a supplement, be sure to select a product that comes from a reputable manufacturer. Most products with DHA derived from fish also include EPA, which is a bonus for your heart health. However, if you are a vegetarian, you can also get your DHA from algal sources. Since fish obtain their DHA from algae, these supplements are just as good as DHA derived from fish (MIDAS used algal DHA), without concerns about environmental toxins or sustainability.

Products come in the form of pills, chews, gels, gummies, and liquids, so if you try one and don't like it, keep trying. You'll also find an increasing array of products on the market—from milk, to yogurt, to baby food, to crackers, bread, and snacks—that have DHA as an additive. These can be a good source of additional DHA.

Other omega-3s are found in plant sources like flaxseed oil, walnuts, and canola oil. You can consider these brain food, since they do improve heart health and thus likely boost oxygen to the brain. However, they don't include DHA, so they should be an addition to your DHA supplement, not a replacement.

DHA is water soluble, so it's generally safe to take, although it does thin blood, so anyone who also takes blood-thinning medication—such as Coumadin—should talk to his or her doctor before taking DHA.

centers across the United States and funded by Martek Biosciences.[11] MIDAS was a randomized, double-blind, placebo-controlled study—the gold standard in scientific clinical trials.

For the study, the research team enrolled 485 healthy adults over the age of fifty-five who were experiencing mild memory problems and gave half the participants 900 milligrams of DHA each day for twenty-four weeks. The remainder of the participants took a placebo. All took cognitive tests at the start of the trial and again at the end of the six-month study period. What researchers found was that both groups had improved their performance, reducing their memory errors, but the DHA group showed remarkable improvement, halving their memory errors over the study period. As expected, the DHA group had doubled their blood level of DHA.

How big was the effect? Interestingly, just being part of the study seemed to offer some benefit. Those in the placebo group improved their performance over the twenty-four weeks, most likely due to the benefits of taking part in a study. But when researchers adjusted their results to control for such an effect, the DHA group still performed better—as if their brains were three years younger than their actual ages.

It's important to note that not all studies published have shown DHA to have late-life brain benefits. But if you look at the studies as a whole, as I did for a paper published in *Nature Clinical Practice Neurology* in 2009, it's clear that most point toward protection against late-life Alzheimer's disease.[12] For that article, I put together a summary illustration that showed how DHA increases blood flow, reduces inflammation, and reduces the collection of proteins in the brain associated with Alzheimer's disease.

Just a few years later, I saw my own diagram included as part of a presentation at a large international conference. The author, Veronica Witte, had recently completed research exploring whether DHA, combined with EPA, would have an impact on the size of the hippocampus—a study I had strongly considered doing myself. Her research showed noticeable growth in the size of the hippocampus in participants who had taken this omega-3 fatty acid combination.[13]

Favorite Flavonoids

Flavonoids are compounds found primarily in plants and are a significant feature of the human diet. There are some four thousand flavonoids that have been identified, but not all have been studied for their brain benefits. Many of those that have been studied have been found to offer benefits for a number of different medical conditions. Some—such as blueberries and spinach—were discovered to have antioxidant and anti-inflammatory properties, and thus find their way onto many a brain-food list.

They're there for good reason. Antioxidants do aid brain health, primarily by ridding the body of free-radical oxygen molecules, which it creates naturally in a process called oxidation. Oxidation occurs when our immune system sends germ-busting cells in to attack an intruder. The cells, called microglia, produce free radicals, which attach to the attacker and kill it. That's all very helpful, and quite necessary, but free radicals can also attach to healthy cells and kill them—an unintended side effect that seems to be more common with aging. Antioxidants offset this effect, with the long-term benefit that fewer healthy cells are killed.

But some flavonoids are more than mere antioxidants; they also increase blood flow to the brain and have been shown to boost BDNF, growing the brain by promoting neurogenesis in the hippocampus. Flavonoids can also contribute to the formation of nitric oxide in the body, which relaxes blood vessels, thereby improving blood flow. Some flavonoids help in other ways, too, by activating the energy source—mitochondria—in our bodies' cells, in much the same way exercise does. One—green tea—has even been shown to promote healthy brain activity.

I have studied many of the flavonoids, and of those that I have examined, the ones I have listed here have the strongest scientific data to support them. That's not to say there aren't others, but since there is little commercial value in an expensive clinical trial, such trials are hard to come by.

Here are some brain-friendly flavonoid-rich foods (presented in alphabetical order):

Apples

Apple skins are an excellent source of quercetin, an up-and-coming flavonoid with a wide range of benefits. Quercetin reduces plaque buildup in the arteries and may reduce the risk of hypertension. But it is most studied for its ability to reduce fatigue and increase endurance. A study in animals showed rats given quercetin voluntarily increased their exercise levels and had greater endurance.[14] When researchers looked at slices of the animals' muscles, they saw higher numbers of mitochondria, which you'll recall is the energy source of cells.

The same team of researchers gave 500 milligrams of quercetin twice a day to participants in a human study.[15] After seven days, the study subjects were able to exercise 13.2 percent longer than a control group. This was a small study and more research needs to be done, so for now my recommendation is to stick to natural sources rather than supplementation. You can add quercetin from other dietary sources, which include citrus fruits, berries, onions, parsley, sage, and tea.

Beets

A good source of nitrites, beets can help increase blood flow to the brain. I recommend eating them once a week, as I do in my own home.

Blueberries

Blueberries are well known for their antioxidant and anti-inflammatory properties. And there's ample evidence that they are, indeed, good for the brain. In one study, published in 2012, researchers at the University of Reading in the United Kingdom fed rats a blueberry-rich diet for seven weeks. Compared to a control group of rats eating a regular diet, those who'd had blueberries added to their diet improved their spatial memories and were able to learn at a much faster rate. As expected, blueberry-fed rats also had higher levels of BDNF in their hippocampi.[16] Other studies have shown that the blueberries increase BDNF in the hippocampus, which we know to be key to growing the brain.

Caffeine

There's good news and bad news about caffeine. First, the good news: caffeine has been shown to increase BDNF in the hippocampus, at least in rats. In one study, researchers at the University at Albany, SUNY, put caffeine to the test by comparing four groups of rats: those eating a high-fat control diet, those eating a high-fat diet plus caffeine, those eating a regular diet with caffeine, and a control group eating a regular diet.[17]

They found that caffeine prevented not only weight gain but also the cognitive decline associated with a high-fat diet. Interestingly, caffeine also prevented or reversed a decrease of BDNF seen in the hippocampi of the rats fed a high-fat diet alone.

Now the bad news: too much caffeine comes with its own problems—notably, jitteriness, stomach discomfort, and high blood pressure. Not only that, a caffeine high can be followed by a distinct low, setting up an endless cycle of highs and lows as you drink coffee to recover, crash, then drink more coffee to recover. There isn't clear science telling us the optimal dose, but based on anecdotal evidence and my own experience, I recommend one cup of coffee in the morning and one cup of herbal tea in the afternoon.

Cocoa

I don't often get to tell my patients to splurge when it comes to food, but I do enjoy recommending cocoa for them. Flavonoid-rich cocoa is tied to a small reduction in blood pressure among people who take it, thanks most likely to its blood-flow improving properties.[18] One product that includes cocoa is dark chocolate, which is delicious and brain boosting. (Look for products that have at least 70 percent cocoa.)

Cocoa's blood-flow attributes likely have a brain-boosting effect, but cocoa may also boost BDNF, though more study needs to be done. Meanwhile, in August 2012 chocolate lovers got some news to savor from the *New England Journal of Medicine:* a study found that nations with high per-capita chocolate consumption produced more Nobel laureates than those with lower chocolate consumption.[19]

Curcumin

Curcumin, a bright yellow powder, is a natural product of the curry spice turmeric and has been shown to have a multitude of beneficial effects, including acting as an anti-inflammatory and increasing blood flow to the brain. In animal studies, curcumin has shown potential for reducing the aggregation of toxic amyloid proteins associated with Alzheimer's disease.[20] In one study, curcumin improved spatial learning and memory in rats with amyloid plaques in their brains.[21]

Several years ago, I was traveling in Japan as a visiting professor and had the pleasure of meeting with Dr. Masahito Yamada, a professor and the chairman of neurology at Kanazawa University Graduate School of Medical Science. Dr. Yamada has extensively studied curcumin. As I toured his lab, I was able to see some of this work in petri dishes containing clumps of amyloid plaques. In the dishes to which curcumin had been added, the plaques had dissolved.

Clearly curcumin is a powerful flavonoid with strong potential for brain health. Some people point to the much lower prevalence of Alzheimer's disease in India, where curcumin is a mainstay of the diet. In India, the rate of Alzheimer's disease is about a quarter the rate in the United States (although, of course, there could be other factors driving that difference).

Unfortunately, curcumin is not easily absorbed into the bloodstream. Researchers are on the hunt for better ways to facilitate absorption of curcumin. (One option is to combine it with piperine.) In the meantime, I recommend adding it to your meals as a spice. My wife, Bita, who takes great pains to ensure our family follows a brain-healthy diet, makes curcumin an addition to many a meal.

Pecans and Pistachios

These nuts can be good sources of choline, an essential nutrient that increases a memory-related neurotransmitter in the brain called acetylcholine. Pistachios also supply vitamin E and omega-3s, so I recommend adding both to your diet as snacks. Like other nuts, walnuts are also rich in antioxidants.[22]

Red Grapes

The skin of red grapes contains resveratrol, a polyphenol with antioxidant properties that has been shown to increase blood flow to the brain and to increase the production of BDNF in animals.[23] Resveratrol protects against heart disease by reducing inflammation and platelet aggregation, which reduces the risk of heart attacks. It is believed to also protect against cancer and diabetes.

Peanuts and mulberries are other sources of resveratrol (and cocoa supplies it in lower doses). The big question is not whether resveratrol helps but how much is needed to experience a brain benefit. For now, I recommend simply adding resveratrol-rich flavonoids to your diet rather than taking a supplement. But there is such strong evidence for resveratrol that I am in the midst of conducting a clinical trial to document its role in improving memory. In that study, my team is comparing resveratrol with Fruitflow (a tomato extract), which seems to have properties similar to resveratrol.

Spinach

Like blueberries, spinach has been shown to be an antioxidant with brain benefits. In one study by researchers at the University of South Florida's College of Medicine, rats fed a diet with spinach supplements for six weeks were able to learn faster than those fed a normal diet.[24]

Spinach is also high in protein, making it a good addition to a balanced diet. (Getting the bulk of your protein from red meat, by comparison, also increases your exposure to fats.) And spinach is a good source of vitamins A, C, K, E, and some B, folate, and calcium. I love spinach and add it to my diet any chance I get.

Tea

Black and green tea are excellent sources of L-theanine, an amino acid that has been shown to enhance the brain's alpha waves[25] as well as increase BDNF.[26] One animal study showed that L-theanine can actually increase neurogenesis in the hippocampus.[27]

Green tea, meanwhile, has other benefits too, thanks to epigallocatechin gallate (EGCG), a polyphenol and antioxidant. One animal study showed EGCG promotes neurogenesis in the hippocampi of adult mice,[28] while another animal study showed it revved up mitochondria in diabetic mice and improved their glucose metabolism.[29] Black and green tea also contain quercetin, another antioxidant.

Tomatoes

Tomatoes are a gold mine of vitamins and antiplatelet ingredients that ease flow in blood vessels, thus increasing oxygen supply to the brain. A recent study by researchers in Finland followed 1,031 men for twelve years and showed that those with the highest blood levels of lycopene, derived from tomatoes, were 55 percent less likely to suffer a stroke than those with the lowest levels of lycopene.[30] Add tomatoes to your diet, and don't forget tomato juice as a smart snack choice.

The Vitamins Your Brain Needs

We know from Dr. Bowman's study that higher levels of vitamins B, C, D, and E in the bloodstream are associated with better cognitive performance and a

Snack Attack

Between meals, snacking can actually offer benefits that go further than merely quieting that tummy growl. A well-timed, carefully selected snack can give you an energy boost and help reduce overeating.

My favorites are:

Nuts: Try a palmful of pistachios, walnuts, or other nuts. One caveat, though, is to watch your portions. Overindulging in nuts can tack unneeded calories onto your daily total.

An Apple: Relatively low in calories (a medium apple has about 81 calories), apples offer vitamins A and C, plus L-theanine, quercetin, and other nutrients.

Cherry Tomatoes: Pop a handful of these (half a cup has just 15 calories) for a sweet, juicy treat. Tomato juice also offers brain benefits.

bigger brain. We also know, from numerous studies, that low levels of certain vitamins are associated with cognitive decline in old age. Key vitamins that affect brain performance are:

Vitamin B12

One of eight B vitamins, B12—along with folate—is vital to the function of the central nervous system as well as the nerves that run to your toes and fingers. Among other things, B12 and folate are important for the formation of the myelin lining that insulates the branches of nerves. When B12 and folate levels fall, this protective lining shrinks. As a result, signals aren't passed as efficiently from neuron to neuron, resulting in reduced cognitive function and slowed thinking. Low B12 levels are also associated with fatigue and depression.

Low levels of these vitamins have another side effect: both B12 and folate are essential for converting the amino acid homocysteine. When B12 and folate levels are too low, homocysteine isn't converted and blood levels of this amino acid rise. The result is an increased risk of heart attack, stroke, and inflammation of the blood vessels. Studies have also found that people with Alzheimer's disease and vascular dementia have high homocysteine levels, compared to their healthy peers, but we don't yet have proof that one causes the other.[31]

We do know that high homocysteine is tied to a smaller hippocampus. One study of 1,077 cognitively healthy elderly people between sixty and ninety years old found that elevated plasma homocysteine levels were associated with shrinkage in the hippocampus.[32] Another study found that high homocysteine levels were associated with a significantly smaller hippocampus and a high rate of hippocampal atrophy over a two-year period.[33] The bottom line is that high homocysteine goes hand in hand with a smaller hippocampus, and since B vitamins and folate are key to keeping homocysteine in check, it's critical to ensure you're getting enough B12 and folate in your diet or through supplementation.

It's important to note, too, that homocysteine levels tend to rise with age, in part because the elderly tend to have a harder time absorbing nutrients. Hence, it becomes harder as you age to get enough B12 and other vitamins through your diet. In fact, as you age you may only absorb 30 to 40 percent of the B12 that you need for optimal health.

Vegetarians and those who've had certain gastrointestinal surgeries need to be particularly careful to ensure they're getting enough of these vital nutrients in their diets or through supplementation.

The good news is that a simple blood test can determine your B12 level, and it's easy to supplement with B12 if needed. For most, that means taking a B12 supplement. For those with severe B12 deficiency, however, monthly shots of B12 may be necessary. Just one note of caution: most laboratories consider a B12 level of more than 200 pg/mL to be normal, although for cognitive health I consider the optimal level to be 500 pg/mL or more. Therefore, you might be told your B12 is normal and still benefit from supplementing or increasing the B12 through dietary changes.

You can add B12 to your diet through dairy products, lentils, spinach, meat, poultry, and fish, which are all good sources. If your B12 level is less than 500 pg/mL you can also take a B12 supplement at a dose of 500 to 1,000 mcg per day.

Vitamin D

Vitamin D is probably best known for its effect on bone health, but it actually is involved in a number of chemical processes in the brain. Low vitamin D has been associated with neurological and psychological problems that range from fatigue and depression to schizophrenia, multiple sclerosis, and dementia. Low vitamin D has also been associated with low BDNF and a thinner cortex.

I check vitamin D levels in all my patients and find that about 20 percent have levels that are low. Fortunately, supplementation works well for increasing your level of vitamin D. Keep in mind, though, that people who have low vitamin D often also have low calcium levels, so simultaneous supplementation for both might be beneficial for overall health.

Vitamin E

An anti-inflammatory substance, vitamin E has many benefits for protection against heart attacks and inflammation in the brain. There's evidence, too, that it may delay the onset and even slow the progression of Alzheimer's disease. Some studies have emerged in recent years, however, to suggest that high doses of vitamin E can be harmful. More research needs to be done, so for now I don't recommend high-dose supplementation for my patients. If you do choose to take a supplement, the ideal dosage is 200 to 400 IU per day, and it's most effective when taken with vitamin C. Many multivitamins include vitamin E, so it's important to check your multivitamin dose before you supplement with vitamin E. People who take Coumadin or other blood thinners shouldn't take vitamin E because it increases the risk of bleeding.

The Jury Is Still Out

There are countless supplements that are touted for their brain-boosting potential but for which the evidence is contradictory or simply not yet available. Two that I often hear about are Ginkgo biloba and coconut oil.

Ginkgo Biloba

Though it's received plenty of hype over the years, the truth about Ginkgo biloba is that, when it comes to science, the evidence is controversial. Some studies have shown it to be beneficial, but two placebo-controlled clinical trials failed to show any cognitive benefit. Since it may have harmful effects as well, I don't recommend it for patients with brain issues.

Coconut Oil

Coconut oil in recent years has gained attention for its potential in helping to treat, and perhaps reverse, some of the effects of Alzheimer's disease. The theory that's been floated is that the brain cells in patients with severe dementia die in part because increased resistance to insulin has impaired their ability to use glucose for fuel (as normally happens in the brain). An alternative source of fuel for brain cells can be ketones; so coconut oil, which is metabolized to ketones, could potentially help fuel the brain when glucose can't.

It may be true, but more study needs to be done. Given that it, too, may have harmful effects, I do not recommend it.

Add a Glass of Wine?

When I give lectures, I find that of all the things I say the one men remember most is this: alcohol in moderation is believed to offer neuroprotective benefits. Often people read that as license to hit the bottle on a regular basis. And maybe it is. Alcohol, after all, is thought to increase HDL, which is beneficial to heart health (and by extension, brain boosting).

But the key here is moderation. As you'll read in chapter 14, alcohol abuse wreaks havoc on the brain. Moderation means one serving of alcohol per day for women and two for men. But you shouldn't consider quaffing a glass of merlot with dinner to be sufficient for a bigger brain. I actually consider it an option that's fairly low on the list of priorities. (A healthy diet, complete with DHA, is right at the top.) That's because if you're gobbling Ho Hos and gallons of soda, ending the day with a glass of merlot will barely make a dent in the damage.

There's an important exception to the moderation rule, too. If you're currently experiencing memory problems, you should avoid alcohol entirely.

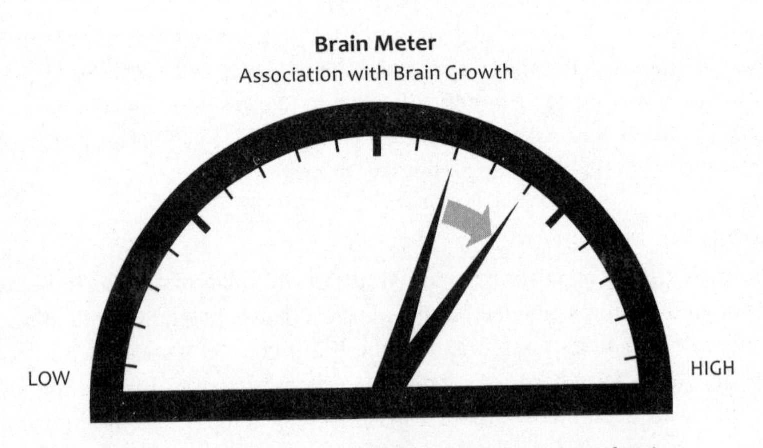

A diet high in DHA, flavonoids, and vitamins and low in trans fats is associated with a bigger brain. The more you adhere to such a diet—and work to avoid being overweight—the bigger the brain benefit. For optimal brain growth, I recommend a diet rich in fruits, vegetables, grains, and fish, (with zero sugary junk food and sodas) along with a DHA supplement.

Brain-Boosting Foods

Food	Good source of
apples	choline, L-theanine, quercetin, vitamins A and C
beets	fiber, folate, nitrites, potassium, vitamin C
blueberries	antioxidant, beta carotene, folate, vitamins A, C, and K
carrots	antioxidant, beta carotene, folate, vitamins A and B6
elderberry	antioxidant, quercetin, vitamin C
leafy greens	beta carotene, calcium, folate, lutein, vitamins A, C, and K
oranges	folate, thiamine, vitamins A and C
pomegranates	antioxidant, folate, vitamins C and K
red grapes	calcium, resveratrol, vitamin C
spinach	antioxidant, choline, folate, protein, vitamins A, B6, and K
sweet potatoes	antioxidant, beta carotene, folate, vitamins A, B6, and C
tomatoes	beta carotene, folate, lutein, lycopene, vitamins A, B6, and C
dairy products	calcium, riboflavin, vitamins B12 and D
eggs	folate, protein, riboflavin, vitamin B12
poultry	iron, niacin, vitamins B6 and B12, zinc
fish	DHA (anchovies, mackerel, salmon, tuna), protein, vitamins B12 and D (some fish)
clams	iron, potassium, vitamin B12, zinc
oysters	iron, vitamins B12 and D, zinc
dry beans and peas	fiber, folate, protein, vitamin K, zinc
chickpeas	fiber, folate, protein, vitamin B6
soybeans	fiber, folate, iron, protein, vitamin C
quinoa	copper, iron, vitamins, zinc
fortified cereal	folate, iron, all the B vitamins, zinc
whole-grain products	folate, iron, niacin, vitamins B6 and E
flaxseed	folate, omega-3, thiamine, vitamin B6
canola oil	omega-3, vitamins E and K
vegetable oil	iron, vitamins E and K
pecans	choline, copper, thiamine, zinc
pistachios	antioxidant, choline, omega-3, vitamin E
walnuts	folate, omega-3, vitamin B6
tea	EGCG (mostly green tea), L-theanine, quercetin, vitamins A and D

Your Rx

Track 1

If you're currently consuming a highly unhealthy diet, with few brain-building nutrients, this is your track. First, you'll need to detox, cutting the worst foods from your diet before you make any serious effort to add brain builders. Why? Your unhealthy diet is increasing your risk of being overweight or obese and of developing high cholesterol, high blood pressure, diabetes, and stroke, all of which are major brain shrinkers. Adding DHA or green tea to your diet won't have nearly the effect that cutting major brain shrinkers will.

Week One: Commit to cutting your consumption of trans fats almost entirely and your consumption of simple carbohydrates—often found in processed foods—by 50 percent. Try to avoid sugary foods, such as donuts, which are high on the glycemic index, as they'll cause a spike in your blood sugar.

Week Two: Continue to stay away from junk food, and now cut down salt and cholesterol. Choose lean meat and eat it no more than once or twice a week.

Week Three: Continue with the cuts you made in weeks one and two and work on reducing serving sizes and limiting caloric intake. Consider tracking your calorie intake for one week through an online tool or app. Seeing just how much a bagel and cream cheese "costs" you in calories can help you convince yourself to tuck into a bowl of yogurt with blueberries instead.

Track 2

If you've already detoxed, or you already eat a somewhat healthy diet, this is your track.

Week One: Focus on eating natural rather than processed or fast food. Be sure to include flavonoids.

Week Two: If you drink coffee, cut back to one cup in the morning and none in the afternoon. Instead, add a cup of tea in the afternoon.

Week Three: Focus on adding salmon or other fish to your diet, and add more fruits and vegetables.

Track 3

If you're already eating a healthy, balanced diet, this is your track.

Week One: Supplement your diet with 1,000 milligrams daily of DHA. (Note: If you take Coumadin or another blood thinner, do not take DHA.)

Be sure you know the levels of your vitamins B12 and D. If they're low, supplement. If they're borderline, adjust your diet to feature more foods that offer vitamins B12 and D.

Week Two: Continue with supplementation. Don't forget to get adequate water. Your brain needs water to function well, so be sure to drink six to eight glasses a day. Add a glass of wine per day several days a week.

Week Three: Continue with supplementation, and continue to eat healthy portions and brain-friendly foods. Try to taste new flavonoids you've never tried before.

CHAPTER SIX

The Path to a Calmer, Sharper Brain

"I WANT YOU TO imagine you're in a field with green grass all around you and a big, beautiful tree in the distance," Dr. Eylem Sahin says in a calm, quiet voice. "Now, walk toward the tree and sit down under it. Relax." She is talking to Beth, a twentysomething patient of mine whose memory and attention issues have landed her in my brain fitness program.

Beth's eyes are closed and she's nestled in a comfortable chair in a dimly lit meditation room in my Brain Center offices. She has already gotten herself into a meditative mind-set by focusing on her breathing and relaxing a succession of body parts—moving from her toes up through her legs to her torso, her shoulders, arms, hands, and then finally her face. By the time she reaches the spot between her eyebrows, she has banished extraneous thoughts—about the paper she has to write for school, or what she'll eat for dinner, or who might be texting her.

Mentally relaxing under the tree, she's halfway through her meditation session, guided by Dr. Sahin, a licensed clinical psychotherapist and the director of my brain fitness program. For the next ten minutes, Beth will nonjudgmentally note whatever thoughts and feelings come to mind. Then, she'll turn her attention again to her body, progressively focusing on relaxation from toe to head.

Beth, who suffers from attention deficit disorder, will complete twelve weeks of these meditation sessions, along with cognitive skills training and neurofeedback (which you'll learn about in a moment). In conversations with our "brain coach," she'll be guided in other areas of her life—exercise, sleep, and diet, to name a few.

All of these efforts are aimed at improving Beth's cognitive function by growing her brain. If we were to hook her up to an EEG after her meditation, we'd likely find her brain operating primarily in the alpha zone, the type of brain activity that reflects a calm, focused, and attentive state of mind. Even better, after a few weeks we'd likely begin to see that pattern even long after she stops meditating. Why? As you'll learn in a moment, mindfulness training promotes healthy brain activity in the short and long term and literally changes the structure of the brain.

You'll recall from chapter 2 that healthy brain wave activity has some parallels to a healthy heartbeat. Just as you'd be alarmed if your heart raced at 120 beats per minute for a prolonged time for no apparent reason, you also want to avoid having your brain "race" in the high beta range (higher than 25 hertz) without good reason. Instead, the ideal is to have overall brain activity in the alpha range of 8 to 12 hertz during periods of alertness—with the exception of short bursts of beta (13 to 25 hertz), which are necessary to perform certain tasks. Alpha activity is associated with calmness, alertness, focus, and attention. It's the zone I'm in when I'm putting my six-year-old to sleep and playing with her hair, and the zone Beth is in as she meditates.

Amazingly, just as we can condition the heart through regular exercise, so that it thrums at a calm 60 beats per minute while at rest, we can also train brain waves to remain in the ideal alpha zone for long stretches of the day. And it's not just meditation that can get us this mindfulness benefit; neurofeedback offers it, as do other activities such as yoga, tai chi, and even prayer.

Om? Growing Your Brain with Meditation

As Beth sat with Dr. Sahin and imagined herself relaxing on the grass under a tree, she was engaging in a type of quiet contemplation called mindfulness meditation. It is just one style among many, each of which has its own methods and strategies.

That broad variety makes definition difficult, but in general the term meditation refers to the practice of deliberate techniques aimed at inducing a state of relaxation, attentional focus, or contemplation. It may involve repeating certain words or phrases (a mantra—"om" is the stereotypical one), regulating your breathing, clearing your thoughts, suspending your logical thought process, or directing your thoughts in a particular way.

Whatever the method, the goal remains largely the same: to train your mind so that you can bring yourself to a state of consciousness that benefits you in some way, be it simple relaxation, heightened awareness, enhanced concentration, or even the achievement of an "enlightened" state of being. Along with the temporary state of relaxation they achieve with every session, meditators often report long-term benefits that spill over into their day-to-day lives, from emotional stability, to improved physical health, to better sleep.

No surprise, then, that meditation has become fairly common in recent years. In 2008, when the National Institutes of Health surveyed Americans to get a handle on how common alternative and complementary medicine practices are in the United States, it found that about 12 percent of Americans say they use deep breathing exercises and 9.4 percent report practicing meditation.

Do You Know How to Breathe? Maybe Not

When Dr. Sahin meets with patients for the first time she often surprises them with a quirky question: "Has anyone ever taught you how to breathe?" she'll ask.

She's not questioning their ability to fill their lungs with air. What she's really after is technique. Proper breathing, as it turns out, is an important part of most, if not all, meditation practices.

Often patients will smile and laugh and then admit that, no, no one has ever taught them to breathe. Dr. Sahin then gives them a crash course in abdominal breathing, which involves expanding your abdomen when you breathe in and contracting it when you breathe out. How do you know if you're doing it right? If you place your right hand on your chest and your left hand on your belly, you'll notice that your left hand moves when you breathe in and out, while your right hand doesn't.

Our patients use abdominal breathing during meditation, but I also have my own favorite simple breathing exercise to get me into a relaxed, peaceful state of mind when I don't have time to meditate. I call it my "7-7-7" technique. To do it, you simply breathe in while counting to 7, hold your breath while counting to 7, and breathe out while counting to 7.

How It Grows

When I tell people meditation is good for the brain, they nod knowingly. "Stress relief!" they may say emphatically. Stress relief is indeed a benefit of meditation, as anyone who has spent ten minutes focusing on their breathing can tell you. If you're deep in a meditative state, by definition you can't be stressed.

That's no small thing. Since chronic excess stress, as you'll read in chapter 10, can release toxic levels of the hormone cortisol into the bloodstream, meditation qualifies as a brain builder just by virtue of reducing stress. By lowering cortisol levels on a daily basis with regular meditation, you're protecting your brain from the toxic effects of excess cortisol, for better brain function in the short term and increased brain growth in the long term.

But meditation does far more than merely quell stress. For starters, as Beth can attest, it is one of the most effective ways to promote healthy brain activity. Meditation, after all, reduces excessive, choppy beta waves and encourages the alpha waves of a calm, focused brain.

Not only that, but meditation helps the various parts of the brain operate in harmony. In much the same way that members of an orchestra play their different parts in sync with each other, the brain's various parts operate at different frequencies but in a complementary way. Imagine if the violins played slower than the expected tempo, while the horns played too fast—the resulting cacophony would be unpleasant, to say the least. You can think of meditation as the conductor who keeps all the brain's parts playing in concert.

In addition to promoting healthy brain activity, we now know that meditation actually increases the size of the hippocampus and boosts BDNF.[1] In one study, researchers looked at the benefits of meditation in patients with stress and found higher BDNF levels in the meditation group. In another review article, researchers found that the decrease in cortisol associated with meditation is linked to higher levels of BDNF.[2] More research needs to be done to assess just how much meditation affects BDNF, but as an expert in the field, I feel confident that once it's done, such research will produce solid evidence that meditation is a BDNF booster, even in people without excessive stress.

Meditation—like other mental training—also strengthens the pathways it "works out," leading to more synapses and greater oxygen flow in those areas (more on this in a moment).

Studies have shown, too, that meditation improves sleep and immune system function and reduces inflammation as well as blood pressure and heart rate, thus improving cardiovascular health and sending more blood to the brain. In fact, one 2010 study conducted by the University of Pennsylvania Medical Center found that experienced meditators had significantly higher blood flow to the prefrontal cortex, parietal cortex, and other areas of the brain that support attention, regulation of emotion, and autonomic function.[3] The study lends support to the notion that meditation results in real biological changes to the brain's function and structure.

Paying Attention Pays Off

Now that we know the mechanisms meditation uses to effect change in the brain, the next question becomes, to what end? For starters, since meditation is primarily the regulation of attention and emotion, it's no surprise that those are two domains of brain function that show the most benefit. In other words, as you practice regulating your attention and emotional response, you improve your ability to pay attention and regulate your emotions.

Some of the most interesting studies showing this come out of the University of Wisconsin–Madison, where Professor of Psychology and Psychiatry Richard Davidson and his colleagues have for years delved into the science of meditation. Davidson has studied meditators of all types—from Tibetan monks to novices. What he's found is compelling evidence that meditation changes the way the brain performs, even in those who don't commit a lifetime to quiet contemplation.

In one study, for example, his research team compared the brain activity of participants practicing a focused-attention type of meditation with those who didn't practice any meditation at all.[4] For the study, participants—both novices and experts—were asked to meditate using an external visual point so that they achieved a "rest condition," at which point they were asked to stop meditating. Researchers then performed fMRIs of their brains. They found that those who'd practiced more showed more activation in the brain regions responsible for paying attention. Expert meditators also appeared to be better able to regulate their emotional responses—measured by activity in their amygdalae—to sounds such as a baby crying.

In another study, Davidson and colleagues linked meditation training and the ability to process information.[5] For this study, the team measured study subjects' ability to process information presented in rapid succession—for example, two numbers embedded in a string of letters. The average person will miss a certain proportion of the second number embedded in the series because his or her brain is too focused on the first. But study subjects who'd participated in three months of intensive training in focused-attention meditation actually outperformed non-meditators and improved their baseline scores. After meditative training they didn't miss the second target as often as they had before. And from scalp EEG monitor readings, researchers could see that meditators were able to complete the test more easily than non-meditators. The results suggested that meditators were able to perform efficiently and at high speed without straining their brains.

Changing Brain Structure

While Davidson and his team were busy in Wisconsin, a research team led by my colleague at Harvard, neurobiology researcher Sara Lazar, was also at work, studying how meditation affects the structure and size of the brain.

Lazar had long been interested in the effects of mental training on the brain. An avid runner, she'd suffered an injury in 1994 that led her to taking up yoga, which combines physical and meditative exercises. Before long, she realized yoga was far more than mere physical therapy. "It had a profound effect on how I view the world," she says now. Yoga, Lazar felt, had changed the way her brain operated. "Being a scientist, I thought, how is this working?"

In 2005, Lazar and colleagues added to the small but growing body of research on the topic with their study of the effects of mindfulness meditation on the brain.[6] For the study, the team compared the brains of fifteen non-meditators with those of twenty people who practiced insight meditation, which involves focusing attention on internal experiences. These were fairly devoted meditators—on average they meditated for about forty minutes a day, although the length of time they'd been practicing meditation ranged from a year to several decades.

When the research team compared the MRIs of meditators with the control group, they found that those who meditated had a thicker cortex in certain parts of the brain, compared to those who hadn't meditated. What's more, meditation appeared to slow the cortical thinning that's a normal part of aging. Incredibly, in one area of the brain the thickness of the cortex of forty- to fifty-year-old meditators was similar to that of twenty- to thirty-year-olds. In other words, portions of their brains actually looked *decades* younger than expected.

Still, there was the inevitable question: Did meditation lead to a thicker cortex, or was having a thicker cortex part of what drew people to meditation in the first place?

Lazar and colleagues aimed to find out. To do so, they enlisted sixteen healthy adults who were enrolled at the University of Massachusetts Medical School's Center for Mindfulness in a mindfulness-based stress reduction (MBSR) course.[7] None had meditated in the prior six months and all had had no more than ten meditation classes in their lifetimes. In other words, they were novices when it came to meditation. They ranged in age from twenty-five to fifty-five years old.

Participants were given MRIs before beginning a program of mindfulness meditation and again after eight weeks of weekly two-and-a-half-hour meetings plus one full-day class. During these periods of instruction, participants were trained in mindfulness strategies, including awareness of present-moment experiences, a "body scan" during which they sequentially focused their attention on sensations affecting various parts of the body, mindful yoga, and sitting meditation.

They also used forty-five-minute recorded, guided mindfulness exercises to practice daily at home. To integrate mindfulness into their daily lives, participants were even told to practice it informally whenever they could—while walking, washing the dishes, folding laundry. Over the eight-week course, the meditating group reported spending about three hours a week on their homework exercises.

The results, published in 2011, reported that the meditators all saw an increase in the size of their hippocampi, plus increases in the size of the posterior cingulate cortex, the temporoparietal junction, and the cerebellum—areas involved in learning and memory, emotional regulation, self-referential processing, and perspective taking. Changes were significant enough to be detected on MRI with the naked eye. No such brain growth was seen in the control group.

Of course, the study size was small and participants were people actively seeking out stress reduction. It's also possible that other aspects of the meditation classes helped spur growth in grey matter, say the study authors. The classes, after all, involved social interaction, stress education, general learning, and even gentle stretching exercises.

Lazar's wasn't the only study to document changes to the brains of meditators, however. A team of researchers led by UCLA's Eileen Luders has offered up the results of its study of twenty-two meditators and twenty-two non-meditators.[8] Those in the meditating group practiced a variety of styles and had been practicing meditation between five and forty-six years, with an average of twenty-four years. Most meditated daily and many included deep concentration as part of their practice.

MRI imaging showed meditators had "significantly larger volumes" in their hippocampi and increased grey matter in the area of the brain associated with controlling mood and drive. It didn't seem to matter what style of meditation the volunteers practiced either. No one style of meditation showed a greater effect than another.

Luders and colleagues also looked at the effect of meditation on brain pathways, this time using diffusion tensor imaging (DTI), for a study published in 2011.[9] Looking at twenty-seven long-term meditators and twenty-seven non-meditators, the team found significantly greater integrity of white matter—the brain's highways—throughout the brains of the meditators, indicating stronger connectivity.

You'll remember from chapter 1 that the integrity of these highways in the brain is vital for optimal brain performance. Keeping the highways running smoothly doesn't just mean messages will pass quickly; it actually helps to keep the neurons alive and strong, helping to maintain and grow the brain.

Try It at Home

I highly recommend meditating under the guidance of an experienced teacher, at least until you've learned a technique. That's not because meditating is hard, but it does take some getting used to. At my Brain Center, we often find that patients have a hard time clearing their minds of extraneous thoughts at first, but after three or four sessions—and a little help from Dr. Sahin—they find it much easier.

There are many different methods, but the one we use at my Brain Center is performed like this:

Start by sitting on a comfortable chair in a quiet room with the lights dimmed. (You may find it enjoyable to add quiet, gentle music with no lyrics.)

Close your eyes. Try to clear your mind and push away extraneous thoughts. Whenever a stray thought comes to mind during the course of your meditating, just push it gently aside. Remember to count slowly at each step.

Breathe in using abdominal breathing and slowly count to three. Breathe out and count to three.

Focus your attention on your **toes**. Imagine a tingling sensation there. Relax your toes.

Breathe in and count to three. Breathe out and count to three.

Focus your attention on your **calves**. Relax your calves.

Breathe in and count to three. Breathe out and count to three.

Focus your attention on your **thighs**. Relax your thighs.

Breathe in and count to three. Breathe out and count to three.

Focus your attention on your **abdomen**. Relax your abdomen.

Breathe in and count to three. Breathe out and count to three.

Focus your attention on your **shoulders**. Relax your shoulders.

Breathe in and count to three. Breathe out and count to three.

Focus your attention on your **neck**. Relax your neck.

Breathe in and count to three. Breathe out and count to three.

Focus your attention on your **chin**. Relax your chin.

Breathe in and count to three. Breathe out and count to three.

Focus your attention on your **cheeks**. Relax your cheeks.

Breathe in and count to three. Breathe out and count to three.

Focus your attention on the spot right **between your eyebrows**. Relax your eyebrows.

Breathe in and count to three. Breathe out and count to three.

Imagine you are in a field with green grass as far as you can see. In the distance is a big, beautiful tree. Imagine yourself walking toward that tree. Sit down under it and relax.

Breathe in and count to three. Breathe out and count to three.

Allow yourself to notice any thoughts, feelings, or sensations that come to mind. Don't analyze them; just note them. You're simply paying attention, not thinking of them as good or bad or trying to think more deeply about them. Do this for five minutes.

Breathe in and count to three. Breathe out and count to three.

Imagine yourself standing up and leaving the tree, walking back to where you started.

Breathe in and count to three. Breathe out and count to three.

Focus your attention on your **toes**. Imagine a tingling sensation there. Relax your toes.

Continue focusing on the parts of your body until you reach the spot between your eyebrows. Once you do, you may either start at your toes again or open your eyes and end your session.

How Much Is Enough?

Davidson, Lazar, Luders, and others are still at work unraveling the mysteries of meditation. One question yet to be answered is just how much—or, rather, how little—meditation we need to experience changes in our brains.

The effect of meditation varies by person—some may experience dramatic improvements while others report more modest gains—but Davidson points to evidence that as little as two weeks of meditation can produce "discernible changes in the brain."

In my practice, I often find that patients enrolled in our brain fitness program report improvements in as few as two weeks, but the strongest results occur after participation in a twelve-week program that includes weekly meditation sessions.

Often patients start haltingly, taking a few sessions to become comfortable meditators. Once they've reached that point, however, they're much more likely to meditate at home and incorporate mindfulness into their daily lives, increasing the time they spend meditating as the weeks go on. By about week four, I typically recommend that patients meditate thirty minutes a day, four days a week. And although many report improvements in mood, state of mind, and even cognitive function very early in the program, I usually find that it takes twenty to thirty sessions to form a habit of meditating—and to experience the substantial benefits that come with sculpting your brain.

Neurofeedback: Harnessing the Power of Alpha

Once a week a fortysomething patient of mine named Robert settles into a comfortable leather chair in a small room at my Brain Center. In front of him is a large flat-screen computer monitor attached to a standard laptop. On one ear is a sensor similar to those used for EKG in heart monitoring. Sensors about half the size of a penny are also placed on various spots on his scalp.

Robert is about to take part in a session of neurofeedback, a type of biofeedback that uses electroencephalography, or EEG, to capture and display brain activity on a computer monitor, with the ultimate aim of helping him reduce the symptoms of a concussion he suffered in a skiing accident.

The sensors applied to Robert's scalp record his brain's activity as he watches a computer display. (The electrode on his ear is merely there to cancel out ambient electrical signals.) When he started the brain fitness program Robert's EEG showed excessive slow activity (theta waves) in the front of his brain and excessive fast activity (beta waves) toward the back—a pattern that's attributable to his concussion.

For his neurofeedback session, though, Robert is here to do more than just have his brain activity recorded. Robert is using regular neurofeedback sessions, under the watchful eyes of our EEG brain trainer Nicole Merrill, to retrain his brain and raise his levels of healthy alpha activity. In addition, like an orchestra playing in sync, after training the various parts of Robert's brain will begin to operate in a harmonious fashion, each playing its own part but in a way that complements the others (a notion author Jim Robbins details in his book *A Symphony in the Brain*[10]). To get there, Robert just needs to practice. With neurofeedback, a passive therapy, he'll do that without even realizing it.

On the display, Robert will see a scene—in this case a landscape of mountains. Robert must simply look at his scenery. A brain computer interface—his brain connecting with the computer—will do the rest. As he watches the screen, the computer will track his brain waves and produce a quiet beep every time it records alpha activity. That beep is a reward of sorts. It tells Robert's brain to continue producing alpha waves. Amazingly, his brain will do just that, with no conscious input from him.

The reward reinforces the behavior until eventually Robert's brain trains itself to operate in the alpha zone even when he's not attached to sensors. This is the simplest version of brain plasticity at work. And it mimics the way our brains work in our everyday lives: rewards shape our behavior. If someone tells you that you look pretty in your pink sweater, for example, you might find yourself reaching for pink the next time you're at the mall. Get a kind word or a smile from your spouse when you bring in the newspaper, and you'll be more inclined to do it again the next day.

For Robert the reward is the quiet beep, although he doesn't realize it on a conscious level. As his brain adjusts its activity in search of more reward, he'll replace excessive beta and theta with alpha activity, gradually, over the course of fifteen to twenty sessions. The result will be that Robert will think more clearly and find it far easier to pay attention.

EEG-based biofeedback has in recent years increasingly shown promise in the treatment of concussion, ADHD, epilepsy, migraine, insomnia, anxiety, and other disorders.[11] The evidence for neurofeedback to treat ADHD is particularly strong; it may reduce symptoms just as effectively, if not better than, stimulant medication. And the results are lasting. Unlike medicine, which wears off, neurofeedback is actually sculpting the brain, changing brain wave activity for the long term.

Neurofeedback as a treatment for certain conditions makes sense, but it's also no surprise that neurofeedback can be used for better creativity, even in relatively healthy brains. Such brain performance enhancement is cutting-edge science, so more study needs to be done. But we do have emerging evidence that neurofeedback aimed at increasing alpha wave activity—especially toward the higher end of the range—in healthy adults may boost cognitive function.

In one study by German researchers, of the fourteen subjects given such neurofeedback training, eleven showed significant improvement in cognitive testing by their fifth session.[12] No such improvement was seen in the control group. It's a small study, to be sure, but it has been supported by other studies that have shown improvements in brain performance following alpha training neurofeedback.

In one, researchers in the United Kingdom enlisted twenty ophthalmic surgeons for a study of neurofeedback aimed at improving dexterity in microsurgery.[13] The subjects were split into a control group and a neurofeedback group, and both groups were given a pre-study assessment of their surgical skills. Their skills were by timed and analyzed by expert consulting surgeons. Participants also rated their own mental state and level of anxiety. The neurofeedback group then received eight thirty-minute EEG sessions. When the surgeons were tested again, the neurofeedback group performed 20 percent better in technique, suturing, and overall time on the task. Participants who'd had neurofeedback also reported less anxiety after EEG training.

In yet another study, researchers in Spain tested sixteen subjects, assigning ten to neurofeedback and six to a control group.[14] The neurofeedback group was given one neurofeedback session per day on each of five consecutive days and a working memory test before the first session and after the last. The neurofeedback group showed significantly more upper alpha activity and did significantly better on the working memory test, each improving his or her memory by ten words compared to a one-word improvement in those in the control group.

Other Ways to Calm Your Brain

Drumming

If you press your ear to the door of my offices on a Sunday morning or a Thursday evening, you won't hear the usual hum of a busy medical practice. Instead, what you'll hear—*bum, batta, bat bat, bum, batta, bat bat*—is the product of six patients and a drum instructor, who are seated in a circle, rapping their bare hands against long, tall djembe drums that rest between their knees. They're almost certainly all smiling, but they're not here just for fun. Instead, they've assembled here for a session of drumming, a therapy that is rapidly emerging as an effective method to promote healthy brain activity.

Why would banging on a drum help your brain? Good question. The science on this is still emerging but it's clear, for one thing, that the repetitive nature of focusing to create a drumbeat promotes healthy brain activity in a way that is similar to the state of mind achieved by meditation.

There are other benefits, too. For one, learning to drum requires brain stimulation; you're building synapses when you work to follow a prescribed pattern or coordinate your beats with others in a group. Although it seems simple, drumming requires focus and attention—on creating your own sound and on the sounds that others are creating, as you attempt to play in concert with them. It's a good way to activate your frontal lobes.

Being part of a group also has benefits; there is a social connection, a sense of "we," among group members that helps to improve mood. And the act of drumming can also offer a stress-relieving physical and creative outlet. Most of my patients even report feeling free or joyful after drumming. Bang rhythmically on a drum for twenty minutes and you'll see what they mean.

Although you can drum with sticks, I recommend a drum that uses your hands instead since the feel of your palm or fingers striking the drum can create a pleasant sensation that you don't get with a drumstick. You can pick up a djembe drum online for anywhere between fifty and several hundred dollars, or you can opt for a cheaper set of bongo drums. You can even skip the drum altogether and simply "drum" your hands against your thighs.

Your best bet is to find a class so you can learn a technique that you then practice at home. In the meantime, practice at home on your own. Just create a rhythm of about seven beats and repeat it. Use your favorite songs as guides.

Tai Chi and Yoga

Tai chi and yoga may change brain wave activity in a way that's similar to meditation. In one study, for example, thirty-eight adults participated in a twenty-minute session that included standing tai chi movements and sitting, lying, and standing yoga poses.[15] Before and after the session, they underwent EEG and math tests. After their sessions, participants showed a pattern of brain activity associated with relaxation and alertness. They also did better on the math test.

Yoga has also been noted to promote alpha activity and decrease cortisol levels. As would be expected, one Japanese study found a strong correlation between lowered cortisol levels and increased alpha activity.[16]

Religion

If you're religious, you might find it comforting to know that praying may be helping you grow your brain. One EEG study of Muslims engaging in prayers, for example, showed prayer-induced alpha activity.[17]

Religion has also been linked to a reduced risk of mortality,[18] likely for several reasons. For starters, being part of the community offers up countless opportunities for social interaction—from weekly services, to study groups, to volunteer efforts, and more. In addition, many people find attending religious services to be a calming, relaxing, or even joyful experience. (If you've ever stood in the pews listening to a gospel choir sing their hearts out you can probably understand why.)

A New Attitude

As beneficial as meditation and other mindfulness practices are, their mind-calming effects can quickly be reversed if you're not careful. You won't get much brain-boosting value from thirty minutes of quiet contemplation if you're pessimistic and irritable the rest of the day, for example.

That's why at my Brain Center Dr. Sahin has created a meditation program that incorporates aspects of cognitive behavioral therapy and other interventions aimed at helping patients find satisfaction, calmness, and happiness in their day-to-day lives.

Part of that effort goes toward helping patients assess their attitude or belief system and the way they respond to stressors in their lives. We do this by encouraging them to conduct an "ABC" assessment of the things that stress or concern them. Patients fill out an ABC sheet as follows:

A for *activating event* (something happens): What was it that triggered the negative feelings?

B for *belief* (I tell myself something): When A occurs, what do I tell myself?

C for *consequence* (I feel something): When I tell myself B, how do I feel?

Here's an example:

A: My car runs out of gas.

B: I tell myself it's my husband's fault because he always borrows my car and fails to fill up the tank.

C: I feel angry at my husband.

Often what stresses us is being confronted with something that is out of line with our belief system. If we can change our belief system to bring it in line with reality, we stand a better chance of diffusing our anger and maintaining a peaceful state of mind. In this example, what if you changed B? What if you recognized that your husband is forgetful about the gas tank and therefore you need to be extra vigilant and check the gas gauge every time you drive? If you could change B, your belief that it's your husband's fault,

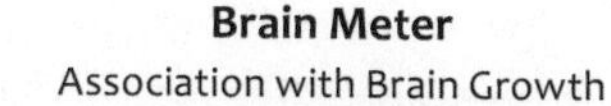

Mindfulness—including stress reduction through meditation or breathing exercises—increases BDNF, enhances cardiovascular health, and promotes healthy brain activity. The more your mind operates in a relaxed and focused state, the more your brain will grow. For optimal brain growth, I recommend meditating (or performing other brain-relaxing exercises) twenty minutes a day, five days a week along with maintaining a positive attitude in life.

you could then likely change C, increasing your chances of maintaining a peaceful state of mind.

Coupled with meditation, the ABC technique helps patients change their attitude toward life, reducing stress and increasing their sense of well-being and satisfaction. The end result is a brain that is more likely to operate in that sought-after alpha zone.

Your Rx

Track 1

If you've never done any type of mindfulness activity, this is your track.

Week One: Start with my simple 7-7-7 breathing exercise. Do it once a day, three times a week, and focus on your technique. If you feel noticeably calmer after doing it, you've probably got it right.

Week Two: Continue with my breathing exercise and add one session of meditation.

Week Three: Shift from breathing exercises to twenty minutes of meditation, four days a week.

Track 2

If you've had some experience in the past with meditation or another form of mindfulness, or are currently taking such a class, this is your track.

Week One: Meditate twenty minutes a day, four days a week.

Week Two: Meditate twenty minutes a day, four days a week, and add a tai chi or yoga class once a week.

Week Three: Continue with meditation and add to your week one more session of tai chi, yoga, or another relaxing activity.

Track 3

If you're already a practiced meditator or regularly engage in some other mindful activity, this track is for you.

Week One: Continue with your mindfulness practices and set aside three half-hour sessions to write down all the things you love about your life. Think about why you love them, and then come up with ways to increase your exposure to the things you cherish. For example, I love spending time with my daughters, Nora and Maya, so one of my thirty-minute sessions might end with a plan that allows me to get home from work a bit earlier twice a week.

Week Two: Continue with your mindfulness practices and set aside three half-hour stretches to write down the things that bother you. Fill out an ABC chart for each and challenge your assumptions about your belief system. Ask yourself, *Why am I unhappy? How can I change my perception?* Come up with creative strategies to change or decrease your exposure to the things that make you unhappy.

Week Three: Continue with your mindfulness practices and start giving yourself a weekly Sabbath. From sunset one day until sunset the following day, promise to do no work. This is your fun time, so fill it with things that you enjoy.

CHAPTER SEVEN

Building Brain "Muscles"

OVER THE COURSE of a few months in the winter and spring of 2012, I spent time every other Thursday evening with a small group of patients who'd enrolled in my brain fitness program. The program, designed to put into action the lifestyle changes we know to have real brain benefits, included educational sessions where I guided my patients on best practices for a bigger brain. Participants also met weekly with our brain coach, who offered tips and encouragement to help them make lasting lifestyle changes. One Thursday a few weeks into the program we sat with our chairs arranged in a circle, ready to practice our signature memorization drill.

"Tonight we're going to memorize twenty-eight random words, in perfect order," I told the group. A month earlier they would have laughed, but on this night each one just nodded. They were ready.

Together we made up the list: book, magazine, TV, water, cat, dog, flower, insanity, mango, beach, sunrise, fear, carpet, airplane, vacation, diamond, astronaut, motor, groundhog, tunnel, igloo, boat, pocketbook, heart, shirt, fertilizer, flashlight, nail.

Then we set to work, using a memory technique that involves breaking information into groups, assigning each group a memorable visual image, and mentally placing that visual image in a location we'd later revisit in order to recall what we'd memorized.

We never left our chairs, but we were soon mentally walking through my office, depositing the items in groups of four at various points along the way. We started in the patient waiting room. "Imagine a six-foot-tall book that takes up half the room," I told them. We pictured ourselves opening the front cover of the book and seeing a cascade of magazines come flowing out toward us. To stop the flow of magazines we threw a TV onto the pile and then imagined water shooting up as the TV hit the magazines. It didn't matter that the image didn't make sense. In fact, nonsense is actually an advantage: visceral images—absurd, sexy, scary, gross, or exciting, for example—are more memorable than the mundane.

With the first four words committed to memory we walked to the next room, the office kitchen, where we imagined a cat chasing a dog around a flower. "That's insanity!" we decided. By now, the group was fully engaged in the exercise and blurting out suggestions for scenarios we'd be sure to remember.

At our next stop, the bathroom, Ned, a sixty-four-year-old real estate agent, took the reins. "Let's imagine there are mangoes in the toilet—that's going to be memorable—and then as we sit down on the toilet we see a poster on the wall with a gorgeous picture of a sunrise at the beach," he said. "Now how do we do fear? Nobody is afraid of the beach." No, "but we might be afraid of flushing the toilet with all those mangoes in it," I said. We all nodded. *That* we could remember.

We continued on through my office. In exam room one, we pictured a lush carpet that covered the floor. It was almost obscured by an airplane, which had somehow been squeezed into the room. We imagined ourselves getting into the airplane to take a vacation and finding a very large diamond blocking our way. By this point, we had committed sixteen items to memory.

The rest of the items we placed in various rooms in my office, making our scenarios as funky and creative as we could.

By the time we'd circled the office and reached the front lobby again, we were ready to recite our list. Ned started us off. Closing his eyes, he mentally retraced our steps. "Book, magazines, TV, water. Cat, dog, flower . . . that's insanity!" he said, laughing. He continued reciting the words, picking up speed as he mentally moved through the office. "This is pretty fun," he said.

"Now, who can do it backwards?" I asked the group. Jennifer, a businesswoman in her late fifties raised her hand and then walked herself through the list in reverse order. "You could have had sixty or eighty words on your list and I think I could do it," said Jennifer.

Before she'd joined the program Jennifer would have scoffed at the suggestion that she could accomplish such a feat. She'd never been one to engage in memory games and didn't consider herself to have a particularly good memory (although, not a terribly bad one either). But once on board, Jennifer had jumped in with both feet, committing to exercise more often, reduce stress, mitigate some key health concerns, and practice stimulating her memory.

And while exercise and eating right were not the easiest changes for her to make, memorization proved to be a simple and fun addition to her daily routine. Driving the same route to work every day, Jennifer started committing names of the cross streets to memory. At the office, she made a concerted effort to remember the name of every new client who came in the door. She was, she told me, shocked at what she could memorize. "It's incredible," she said during a later office visit. "I've always had a decent memory and never had difficulty in school, but I wish I had had this ability way back when. I can't imagine what I would have been able to do!"

Jennifer's mastery of simple memory techniques could be brushed aside as nothing more than a party trick. But what about the fact that she was using them every day? Having a technique made her more successful at memorizing–and more apt to keep at it. She told me she found the challenge fun. But her efforts were more than mere entertainment. By engaging in regular mental practice, Jennifer was creating new synapses in her brain. With vigorous, intense memory practice, she could even *grow* her brain, especially her hippocampus.

How Your Brain Grows

We've long known that practicing a given activity strengthens the connections in the brain regions associated with that activity. But only in recent years have researchers provided solid evidence that specific interventions can selectively strengthen and enlarge different parts of the brain, so much so that improvements can be seen on MRI with the naked eye.

In a medical paper published in the *Journal of Neuroscience,* Emma G. Duerden and Danièle Laverdure-Dupont summed it up like this: "Practice makes cortex."[1] This is particularly true for the hippocampus–practicing memorizing things leads to a bigger hippocampus, for example–but also across the cortex, as you'll read in a moment. That your performance improves with practice is no coincidence: you get better precisely *because* you've added and strengthened synapses, neurons, and fiber bundles.

Even better, if you're learning something new, or improving your ability in a cognitive skill by persistently practicing it, you are literally reshaping the part of your brain responsible for that mental task *and* improving the way it communicates with other parts of the brain.

Scientists are still uncovering just what underlies such incredible growth. We know, though, that cognitive stimulation increases blood flow to the brain. In fact, studies using PET scans have shown increased activity in different parts of the brain when those areas are activated by performing certain tasks. What do PET scans measure? Changes in oxygen flow.

We know, too, that cognitive stimulation is associated with increased levels of BDNF throughout the brain, promoting the survival of new neurons in the hippocampus and aiding in the expansion of synapses elsewhere. BDNF, in fact, is a critical protein in the formation of long-term memory. The more BDNF we can generate, the better will be our ability to remember.

Cognitive stimulation may also change the structure of the brain through some other mechanism. One possibility is that such brain training promotes healthy brain wave activity or helps generate new blood vessel branches, although definitive studies have yet to be done in this arena.

A Growing Hippocampus

Some of the most dramatic evidence of the brain-growing effects of cognitive stimulation—not too surprisingly—centers on the highly malleable hippocampus, the area central to learning and memory. "Practice" leads to a hippocampus that's measurably larger.

There's almost overwhelming proof of this phenomenon. One of my favorite examples is a group of language learners studied by researchers at Lund University in Sweden.[2] The students were members of the Swedish Armed Forces Interpreter Academy and were engaged in a three-month program to learn a new foreign language. The program was intensive: students studied from morning to night, seven days a week. For the study, the research team enrolled a control group of students attending regular college courses. Both groups were given brain MRIs at the beginning and end of the three-month period.

Looking at images of the brains of language learners after three months of intensive study, researchers saw measurable growth in the hippocampus and in areas of the cortex related to language. Even more interesting, not only did the growth occur, but those who learned more also saw their hippocampi grow more. And those who had worked harder to fluently speak the language saw greater growth in the motor region of the cortex, which is tied to moving the mouth. Why? Students practicing a new language must repeat new words, moving their mouths and jaws repeatedly in just the right way for proper pronunciation.

Other studies offer more proof of the long-term brain-growing effects of intensive learning. In one, researchers in Germany set out to determine if intense studying would result in structural changes in the brain. The research team performed brain scans on thirty-eight medical students three months before their national medical exams—called the *Physikum*—again a day after the exam, and finally three months later.[3]

Looking at the MRIs taken immediately after the exams, the research team could see that the students' hippocampi were indeed larger. But even more interesting was what they saw on the MRIs taken three months later. Although the students had stopped studying when they finished their exams, their hippocampi continued to grow. Such growth surprised even the research team. We don't know for sure why it happened, but one possible explanation is that heavy studying may have helped new neurons to be mature enough to make connections to other neurons and thus survive as part of a neural network. Once they reached a critical survival point, the newly born neurons were able to continue to make connections despite the fact that the catalyst–studying–was no longer present.

Another fascinating study, this one of London taxi drivers, offers up more evidence of the power of mental stimulation. In this study, researchers at the University College London examined MRI scans of the brains of London taxi drivers and found that they had larger posterior hippocampi–the part of the hippocampus most closely involved with learning to navigate–than non-taxi drivers. In addition, more experienced taxi drivers had larger hippocampi than their less experienced peers.[4]

Did the nature of taxi drivers' work cause the hippocampus to grow? Or were people who became taxi drivers just more likely to have larger hippocampi, and thus were more likely to stick it out as taxi drivers? The research team put that chicken-or-egg question to the test with another study, this one published in 2011.[5]

This time the team enlisted seventy-nine men hoping to become taxi drivers, plus thirty-one others to act as a control group. The study subjects underwent MRI scans and cognitive testing and then the taxi-driver group embarked on their mission of becoming licensed taxi drivers. To do so in London, you must pass a difficult test called "The Knowledge," which requires memorizing the complex layout of the city's twenty-five thousand or so streets. It's no small feat: taxi driver hopefuls often spend three to four years studying for "The Knowledge" (and yet only 50 percent pass).

From the first round of MRIs, taken before the men started studying for the taxi exam, researchers determined that hippocampal size was roughly the same for everyone–those in the group who wanted to become taxi drivers as well as those who did not. Memory tests also failed to note a significant difference between participants in the two groups.

Three to four years after they'd enrolled, the study participants underwent another round of MRIs and testing. Of the seventy-nine taxi-driver trainees, thirty-nine had passed "The Knowledge" and the rest either failed the test or dropped out of training.

So, what did their brains look like? In trainees who successfully qualified, researchers reported an increase in volume in the posterior hippocampus. In those who failed, no such increase was seen. The study authors had their answer: learning the information required to pass "The Knowledge"–rather than merely training for the test or some other factor–seems clearly related to visible growth in the posterior hippocampus. In short, successful drivers had buffed up their hippocampi, to such a degree that the change could be seen on an MRI.

Learning takes many forms, of course. People who are blind, for example, spend their days learning to navigate the world without the benefit of sight. This is perhaps the reason they have hippocampi that are larger—8.5 percent larger in one study—than their sighted peers. In that study, blind adults learned to navigate a maze faster than people who can see but were blindfolded for the task, proving their life-long "practice" had brain benefits.[6]

And Across the Brain . . .

Of course, the hippocampus isn't the only part of the brain that grows with use. In fact, engaging in certain activities can lead to actual increases in grey matter volume in the frontal, parietal, and temporal lobes. In 2004, researchers published a study that showed the result of training—in this case, training to juggle three balls.[7] For the study, the research team divided a group of twenty-four young adults into two groups and taught one group to juggle. After three months—when the juggling group had become proficient—brain scans were taken and compared to scans taken at the start of the experiment. The juggling group then stopped juggling for three months and another set of scans were taken. By this time, incidentally, most of the jugglers had forgotten how to juggle.

Sophisticated brain imaging showed that although all the participants had similar brain scans at the start of the experiment, after three months the jugglers had more volume in the parts of the parietal cortex important for sensory motor integration. Three months later, with juggling practice and learning stopped, grey matter volume decreased. All along, the non-jugglers showed no changes in their brains. In other words, the parts of the brain responsible for the movements needed to juggle grew with use and shrunk with disuse.

Other studies have shown similar results in people who become highly proficient in other activities. Basketball players, for example, experience growth in the parts of the cerebellum associated with eye–hand coordination and balance, mathematicians' parietal lobes are larger than their peers, and novice golfers experience an increase in the sensorimotor areas of their cortices as they learn to golf. In fact, in the case of the golfers, a visible increase in brain size was noted on MRIs after just forty hours of training in the sport.[8]

If You Use It, You Won't Lose It

It probably doesn't surprise you to hear that cognitive stimulation throughout life can help stave off late-life dementia. Anecdotal evidence has long existed. There's the ninety-year-old who continues to go to work every day and credits his hours in the office with keeping him sharp. Or the eightysomething who says daily crossword puzzles have kept her mentally nimble.

Not all intellectually stimulated people avoid dementia, but research supports the notion that staying mentally engaged—playing board games, reading, playing a musical instrument, or dancing—is associated with a reduced risk of developing dementia and improves cognitive function late in life.[9] We now know people engaging in such activity are protected against Alzheimer's disease because they've built brain reserve throughout their lives, making their brains bigger and stronger.

Perhaps one of the more interesting studies on this topic, though, is being done by my colleagues at Johns Hopkins University, in Baltimore. This research grew out of a program called Experience Corps, which got its start in 1988 when former Secretary of Health, Education, and Welfare John Gardner drafted a concept paper on his idea to pair senior citizens with grade school children—for the betterment of both. The idea was simple: put to use the wisdom, time, and talent of older adults in order to help children at a critical juncture in their educational lives.

The elderly would volunteer in schools, primarily in low-income areas. For the kids, having one-on-one time with an adult who could act as both mentor and tutor would surely lead to higher achievement in school. For the adults, making a commitment to get up and out to "work" several times a week would increase physical activity and social interaction, not to mention offer the psychological reward of helping others and putting to use the knowledge they'd acquired over a lifetime. It would be, Gardner believed, a win-win.

Gardner's plan led to a pilot program that launched in 1996 in five U.S. cities. Volunteers had to commit to being in school fifteen hours a week and spend their time with the same teacher and students for the duration of the school year. The pilot was a success, and by 1998 it had been expanded to other cities.

From the start, the pilot program had included a team of researchers at Johns Hopkins's School of Medicine and Bloomberg School of Public Health, including my colleague Dr. Michelle Carlson. The team's job was to help design the program and then to study its impact.

When it came to the students, Experience Corps was clearly a win. Third-graders in the schools with Experience Corps volunteers scored significantly higher on a standardized reading test than children in the control groups; plus, they had fewer behavior problems. Volunteers, meanwhile, reported being more socially and physically active. And there was something else. "I got a lot of feedback saying 'this really removed the cobwebs from my brain,'" recalls Dr. Carlson, an energetic and inquisitive researcher who is also associate director of Johns Hopkins's Center on Aging and Health. "They were saying it without any prompting from me."

The elderly study subjects weren't imagining it either. When their cognitive skills were tested, those who'd taken part in Experience Corps exhibited improvements in executive function and memory over the study period.[10] In fact, those who'd started the program with the poorest executive function showed gains of up to 51 percent in executive function and memory after six months. Members of the control group who were similarly impaired at the start of the study actually showed declines in decision making (which you'd expect, given the normal course of aging).

But what exactly was going on in the brains of those who'd volunteered? To find out, Dr. Carlson and a group of colleagues recruited eight volunteers to take part in Experience Corps and nine age-matched controls who were waitlisted to take part the following year. The volunteers underwent fMRI scans before being placed in a school. Then they went through training and got to work tutoring and mentoring students in kindergarten through the third grade.

After six months, a new round of fMRIs offered proof that experience significantly changes the brain: those who participated in Experience Corps showed measurably increased activity in their left prefrontal cortex and anterior cingulate cortex as they performed a cognitive test.[11] The Experience Corps volunteers also did better on tests of executive function.

Why? Taking part in the program required cognitive stimulation. Planning how to get to school every day, for example, kicked the frontal lobes into gear, as did working as part of a team of volunteers in order to complete tasks and adapting to the needs and abilities of children in the classroom. The program also offered a "workout" for memory: volunteers had to learn or relearn the Dewey decimal system so that they could lend a hand in school libraries. And

Building Brain Reserve One Thought at a Time

I'm going to give you guidance at the end of this chapter on different types of cognitive stimulation you can try. But while you need to take part in scheduled activities, just as you do for exercise, I also advise making mental gymnastics a part of your every day.

Dining out and ready to sign your credit card slip? Instead of whipping out your phone to calculate the tip, do the calculation by hand or in your head. Make a practice of remembering phone numbers rather than just programming them into your phone. Use your GPS, but read a map too. Predict the total of your grocery bill before the cashier rings it all up. Think! Whenever you can, think.

they had to familiarize themselves with learning materials and remember the names and faces of the students and teachers they interacted with. Finally, being part of Experience Corps offered regular opportunities for social interaction, which requires the use of various cognitive abilities (as you'll soon read).

That they'd offset at least some of the effects of aging by taking part in Experience Corps offers a compelling argument that cognitive stimulation is a worthwhile pursuit at any age. This is tremendously exciting news for everyone but particularly for anyone past midlife who wants to sharpen his or her mind—and keep it sharp longer. It's the reason I make cognitive stimulation a critical part of my brain fitness program, and why you'll need to make it part of your plan to grow your brain. Practice *does* make cortex.

Practice . . . but What?

If you're like most of my patients, your next question will be a logical one: What *kind* of practice? People don't usually stick with activities they don't enjoy, so I always let them pick their own preferred pursuits when it comes to cognitive training. But I do offer some guidelines.

For starters, I recommend choosing an activity that allows you to cross-train different parts of your brain. Carpentry is a good example. If you're building furniture, you'll need to engage your attention and focus to figure out what materials you'll need, what pieces you'll create, and how they'll fit together. You'll use the parts of your brain responsible for mathematical calculations as you measure and calculate the sizes of each piece. And as you operate tools and manipulate the wood, you'll rely on the parts of your brain responsible for manual dexterity, organization, and attention. All told, you've enlisted your frontal, parietal, and occipital lobes. Not bad for manual labor!

I also recommend regularly reaching outside of your routine to add new activities to your life. If you've always played bridge, you can consider it cognitive exercise (it is!), but you won't get as much value for those three hours a week you play bridge as you will doing three hours of something new. Why? Learning something new is a fantastic brain grower, because it actually requires the development of new synapses, rather than merely the strengthening of existing synapses.

Finally, in addition to cross-training within the brain, I recommend cross-training your cognitive stimulation with exercise and social interaction. This multiplies the reward. To understand why, consider the benefit you get from riding a bicycle. It's good exercise, which will help grow your brain. But joining a cycling group is even better, as it adds social interactions that work the mind. Even more bang for your buck: join a cycling group and take on the responsibility for planning the group's routes. Your mapping efforts will give your brain a workout before you even climb aboard your bike, and interactions with your cycling friends add a brain benefit as well.

The Power of Engagement

Exercise and overt cognitive stimulation might seem like no-brainers when it comes to cross-training your brain, but the value of social interaction shouldn't be underestimated either. In fact, it's likely that social interactions are a vital part of the benefit recorded in some cognitive stimulation studies. In the Experience Corps program, for example, it's difficult to tease out how much of the benefit came simply from being regularly engaged with other people. Dr. Carlson thinks the social aspect surely had an impact—and I agree.

Why would regularly engaging with people affect your brain structure and size? You may not realize it, but interacting with others actually requires a good deal of mental maneuvering. To understand why, imagine yourself standing amid friends at a cocktail party. As you mingle and chat, you may need to keep track of several conversations as well as information about those you encounter, making assessments about your relationships and tailoring your conversation appropriately. Remembering that the person to your right is a strong gun rights advocate, and the person to your left isn't, you might decide to steer clear of the topic (or not, if you're interested in a debate). Bantering with someone you don't know well, you might resist the urge to tell a political joke—a sign your frontal lobes are in play.

Such mental gymnastics help us exercise various parts of our brains, building synapses, but they also likely aid brain growth in other ways. Interacting with others often brings us pleasure, which may lead to the release of endorphins or other neuropeptides, similar to the runner's high we get from a long run. Taking part in social activities may also reduce our levels of stress and therefore reduce excess cortisol, which, as you'll soon read, is a major brain shrinker. How much, we don't yet know, but I consider social interaction to be an important component of cross-training the brain.

In addition to engaging in cognitive activities that cross-train, I recommend something else for all my patients: regular memory workouts to grow the hippocampus and improve skills of memorization. That's because the hippocampus, you'll remember, is one of the first areas of the brain to deteriorate as we age. Building it up will help offset that natural process, for benefits now and later. When I suggest memory training, though, I'm often met with a knee-jerk response: "But I have a terrible memory," my patients will protest.

They're not alone in thinking that. Easily 90 percent of my patients feel the same way. It is, in fact, the most common complaint I hear from patients (and from friends at cocktail parties). They are often shocked to learn that even people who perform tremendous feats of memory likely once felt the same way. It's at that point that I tell them about my remarkable friend, Nelson Dellis.

A Champion's Brain

On a crisp spring day near the end of March 2012, I strode onto a stage in the stately nineteenth-floor event room of Manhattan's Con Edison building to face an audience of several hundred people. Though they varied greatly in age, race, and occupation, everyone in the room had one thing in common: they all had memory on their minds. They were here, after all, to witness—or participate in—the fifteenth annual USA Memory Championship, a competition designed to test incredible feats of recall. I was about to launch into a presentation detailing ways to increase brain size and improve memory too. For simplicity's sake, I gave the audience just three pieces of advice: eat well, exercise, and make that brain of yours *work*.

I can't get inside the minds of anyone in the room that day but I'm fairly sure that, at least for a certain subset of my audience, the admonition to put their brains to work didn't come as much of a surprise. After all, they'd spent the morning doing just that, committing to memory the names of strangers, the order of a deck of cards, and long strings of random numbers, among other things.

The competition had kicked off at 8:45 A.M. with the fifty-six participants—or as the competition founders like to say, "mental athletes"—spending fifteen minutes memorizing a series of names and faces. In the front row was Nelson Dellis, an affable six-foot-six twenty-eight-year-old computer scientist and reigning USA Memory Champion from Miami. Not far away, sporting a cowboy hat, was thirty-eight-year-old Texan Ron White, a two-time memory champ whose southern drawl and quick wit would liven up the proceedings throughout the day. Off to one side of the room sat Michael Glantz, an amiable eighteen-year-old who, despite his youth, had managed in 2011 to set a record in the poetry contest.

The names-and-faces challenge was followed by "speed numbers," a test of participants' ability to remember randomly generated digits in five minutes. By 10:45 A.M., the group had moved on to the especially challenging poetry competition, in which participants were asked to memorize a previously unpublished poem.

By the time my presentation rolled around, they'd also completed two rounds of "speed cards," memorizing as many cards as they could in five minutes. As we dispersed for lunch, event organizer Tony Dottino read off the results of the morning's contests. Dellis had come out ahead in the names-and-faces and poetry contests and had even set a record in speed numbers, reeling off 303 random digits in perfect order. In speed cards, eight mental athletes—Dellis included—had memorized an entire deck of cards in five minutes or less.

Rather amazing. And yet, if you'd gone around the room and asked, it's likely that none of the competitors would claim to have been born with memories that were particularly noteworthy. Dellis says his memory before starting to train as a mental athlete was "a bit below average." In part, he blames his once-poor recall on his education in physics, a discipline that requires intense focus on a limited area of interest. "I would be so focused on that one thing that other things just flew into my head and out the other ear, so I was never particularly good at remembering anything, not even numbers," says Dellis. He's being modest, of course, but there is truth to what he says.

In fact, I have no doubt that nearly all the mental athletes competing that day had fairly average memories before they started their training. And most—like nearly everyone I meet—probably mistakenly believed memory to be an innate quality rather than something that can be taught. Good memory or bad, you either have it or you don't, right?

Wrong, actually. As Dellis and other memory champs have proven, memory is a skill that can be vastly improved with practice. Before he stumbled upon an article about the memory championships in 2008, Dellis hadn't really considered that he could one day have a record-breaking memory. But listening to those who were already winning memory competitions convinced him to try it. "They were all saying the same thing, which was that anybody can do this," says Dellis, who bought an audio book on the topic and started honing his memory skills. Almost instantly, he says, "I saw how amazing it worked."

Dellis might have launched his bid for the championships right then but an employment offer shifted him back into workaday mode, with no time to seriously train. It wasn't until 2009, and the death of his grandmother from Alzheimer's disease, that Dellis tackled the task with vigor. Suddenly, having a fantastic memory seemed more important than ever. "That's when I started really training every day," says Dellis, whose laid-back style belies an intensely competitive streak.

That year he practiced one to two hours a day. By the time I met him in 2012 he was up to three to five hours a day. "Not because it takes that long to learn these techniques or to get good at them," he explains. "It's just, to win I had to make sure that I could do it very well." An avid mountaineer—on the day of the competition he sported a T-shirt with the simple pronouncement, "Everest 2013"—Dellis stuck to an impressively rigorous exercise routine that included running and weight lifting, plus outdoor activities. He also worked on his diet, ditching junk food and focusing on healthy nutrition.

The training, of course, paid off. After I ceded the stage on that day in March, Dellis prevailed in the championship rounds—in part by memorizing two full decks of cards—and was crowned U.S.A. Memory Champion. But Dellis says he's reaping more rewards than just a sleek trophy or the bragging rights that come with the title. "I didn't really realize it back then but before I started training I was so sluggish," he says. "Of course my memory has improved but there are a lot of bigger things in my head that have improved. I feel more on point with everything I do."

You're no doubt wondering, by this point, how he did it. There are actually a host of different techniques mnemonists use to overcome the limits we normally face when attempting to remember strings of items, such as numbers, names, or the order of a deck of cards. When he broke the record of the most digits recalled, Dellis used a technique that involves pairing numbers with images in his mind of people, actions, and objects.

To use the technique, Dellis first memorized a long list of numbers that he randomly assigned to people, objects, and actions. For example, 111 might be George Bush, 52 might be swimming, and 95 might be a garden hose. Committing the pairings to memory ahead of time, Dellis is able to use them when presented with any list of numbers, no matter the order. He simply "chunks" the list of numbers into strings of seven and then pictures in his mind the person assigned to the first three, the action assigned to the second two, and the object assigned to the last two numbers.

But there's one more important piece to this technique: the "memory palace," a concept fascinatingly detailed by reporter Joshua Foer, who transformed himself into a mental athlete and memory champion and detailed the effort in his book *Moonwalking with Einstein*. The memory palace is simply a physical location so familiar to you that you can walk through it in your mind with ease. To memorize numbers, Dellis combines his person, object, and action technique with the memory palace, mentally walking through a location and depositing the memorized items as he goes.

When he's ready to recall the items, he simply walks through the location again "picking up" the images as he goes. It's the "walking through" a scene in our minds, of course, that helps aid our memories. Instead of picking images out of thin air, we're finding them in a given location, right where we left them.

For his win that day in March, for example, Dellis translated 0093495 as follows:

009 = Olivia Newton-John

34 = dunking

95 = a helmet

Putting them together, Dellis envisioned Olivia Newton-John performing a slam dunk with a helmet. He placed that image in the first location in his memory palace, which in this case happened to be the porch of his house.

To remember the next string of numbers, 7790141, Dellis created an image of soccer star Steven Gerrard (779) swinging an ax (01) through a sheet of paper (41). That image he left at the foot of the stairs. He proceeded to remember his long strings of numbers and deposited them around the home until he'd committed to memory a whopping 303 numbers.

Dellis thinks he still has room for improvement—and I agree. Not long after the competition he was back at work, this time enhancing his number system so he could remember strings of eight at a time and hoping to improve his memory of a deck of cards with just under sixty seconds to less than thirty seconds.

Can he do it? With enough practice, yes, I believe he can. Could you do it? Yes, provided you devoted the same time and attention to it. After just a few minutes of training with Dellis—and practice on my own—I'm now able to memorize a deck of cards, a task I would not have considered attempting just a few years ago.

Your Memory In Action

You're off to your niece's fifth birthday party, a family barbeque held in her parents' backyard. When you arrive, you're greeted with a hug from the birthday girl and you comment on her adorable pink dress. You follow her to the backyard and wave hello to friends as you go. As you walk, you can smell burgers cooking on the grill and hear the squeals of kids playing tag.

Each of these different sensory experiences is filed in a different place in your brain, without you ever thinking about it. Later you may retrieve all or part of them, intentionally or involuntarily. The smell of burgers, for example, might one day bring you back to that afternoon in the yard. But unless something really noteworthy happened—a clown jumped out of a three-foot-tall cake, for example—you might retain no more than an impression of the day or a fleeting fragment of a memory. If something memorable did happen, you'll likely retain more of the details of that afternoon. That's because your hippocampus will have decided that, of all the information you receive each day, *that* information was worthy of remembering.

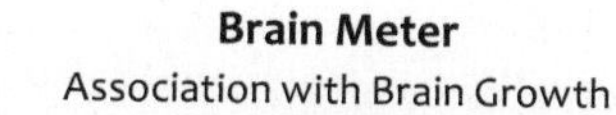

Cognitive stimulation is associated with a boost in the size of hippocampus and cortex. The more you practice memorization skills, take on challenging new hobbies, and learn new information, the more you will build a bigger brain. For optimal brain growth, I recommend fifteen to twenty minutes of memorization practice five days a week, plus one hour a week of a brain super challenge (such as memorizing a deck of cards).

What Makes Us Remember?

You will soon incorporate memory practice into your daily life as part of your twelve-week plan to grow your brain. But before you do, it's helpful to understand some factors that affect the making of memories.

Information that's tied to emotion, for starters, gets etched more deeply than information not tied to emotion, and thus is more likely to be stored long term. So, if you watched news coverage of a bridge collapse or other tragedy, for example, and felt helpless or deeply sad about the situation, you're more likely to remember it than, say, a less visceral news item you saw on the same day.

Your hippocampus also remembers information it deems particularly relevant to you or particularly unusual. That makes sense, if you think of it in terms of evolution. A deer walking in the woods can't pay too much attention to every squirrel it sees. It wouldn't be able to function. But the deer *must* take note of predators for its own survival. So while a squirrel might not register, a strange shadow or the sound of a growl might. For early humans, survival often boiled down to basic matters, such as getting food, procreating, or finding shelter. Today, survival may mean a wide range of things, from success in business, to social skills, to complex decision making in political situations. What we deem memorable, then, has changed as well.

Thinking in the Digital Age

If you sometimes find it hard to think clearly, join the club. In the digital age, when we're bombarded with information at every turn, many people find it extremely difficult to avoid distractions.

Most of us are just victims of too many stimuli. With so much vying for our attention, we've become adept at skimming and rusty when it comes to reading deeply, listening intently, and staying focused. What's more, our wired lives make memorization a relic of a distant day. We're rarely required to commit a phone number to memory or hold driving directions in our heads. And why bother remembering facts and figures when we have access to the Internet from the phones we carry with us everywhere?

In part, it may be a fair trade. I, for one, wouldn't want to revert to the days I had to drive to a medical library and flip through paper journals to research other experts' work. But we still don't know what the ultimate cost will be. I suspect we'll find that the hippocampus suffers as a result of this shift brought on by technology. That's because to skim material you primarily rely on your frontal lobes, not your hippocampus. In-depth learning, on the other hand, requires the work of the frontal lobes *and* the hippocampus. As we skim more and memorize less, we'll likely find our hippocampi shrinking.

That's alarming, for reasons you already know. Having a healthy hippocampus is critical for enhanced brain function today and reducing the risk of dementia tomorrow—one more argument for spending a little time each day reading deeply or practicing memorization.

Of course, not all the information your hippocampus receives will be stored in a retrievable way. Some will be held only for a short time in your memory. If you're making a plan to meet someone next month, you will remember the date and time for just long enough to enter it into your calendar. Unless it's really critical to your success in some way, once it's written down, you'll promptly forget it.

Some information is worth keeping; most is not. You may remember quite clearly the sights and sounds of your wedding day, but you probably won't remember much about a day three months before or after you walked down the aisle. Why not? It wasn't that your hippocampus took the day off. It received plenty of information on those days, too. But it channeled to the cortex only those things it deemed worthy of the effort. Hence, the image of your spouse standing before you and saying "I do" likely will make it to your cortex for long-term storage, but the image of you sitting in your car at a stoplight three

The Technology Gap?

Why is it that downloading an app on an iPhone is so easy that my eight-year-old daughter Nora can do it but so hard that her grandfather would find it too puzzling to master?

It's not that older brains aren't able to understand technology, although that's a common misconception. In fact, difficulty with technology—whether it's surfing the web, creating a spreadsheet, or programming a DVR—has more to do with declines in learning speed than the actual technology itself. As we age, atrophy in the frontal lobes makes learning certain new tasks harder. Older people may have to work more than their younger peers to learn to cook a complex meal or navigate in a new city. But, as you'll remember from chapter 2, their absolute ability typically doesn't diminish. With time, and persistence, a cognitively healthy older person should be able to master any new task.

weeks before the big day will not (unless, of course, your car is rear-ended by a semi at that moment and you're rushed to the hospital, all of which would likely make the long-term memory cut).

Tips to Remember

On the one hand, it's a blessing that your hippocampus handily decides for you which memories to keep and which to discard. Imagine if you remembered every little detail of every experience? You'd have no way to prioritize, no way to sift through all those memories and give them relevance.

Jill Price, who suffered from this rare condition and wrote the 2008 book *The Woman Who Can't Forget,* told ABC News's Diane Sawyer that the experience was like watching a split-screen TV, with the present on one side and memories of the past running continuously on the other. The overload of information left her feeling paralyzed and depressed.

Fortunately, the vast majority of people have no such problem. Their hippocampi work quite well, remembering consequential information and tossing away the rest. Of course, as impressive as it is, the hippocampus isn't perfect. It may fail to flag something you later wish it had. Or send to storage something you'd rather forget.

The good news is that there are ways to influence the hippocampus into sending for storage—and reinforcing in your memory—those things you'd really like to keep.

Keep reading for some suggestions.

Make It Memorable

Your hippocampus will be more likely to send something for long-term storage if it's, well, memorable. Making it so may require converting the information you hope to remember into something more likely to stick out in your mind.

So, what sticks out? Not names, random words, or numbers—that's for certain. But images do, so converting less memorable forms into images is one way to elicit the "save" message. And not just any image, but one that stands out—by being sexy, funny, or absurd. It's the equivalent of tricking that deer into thinking the squirrel is really a wolf.

Even better, give the images a story. By creating a story composed of vibrant, provocative images of the items you're trying to remember, you'll make it far more likely to be stored and more easily recalled. So, if you're trying to remember your grocery list, instead of memorizing each item—pretzels, milk, napkins—you might imagine a giant comic-strip-style pretzel diving into a huge bucket of milk and drying off with a napkin. You've just tricked your hippocampus into saving random words it might otherwise have ignored.

Get Emotional

No, you don't have to tear up every time you want to remember an acquaintance's name, but since your brain is more likely to remember something that's tied to an emotion, you can trick it into doing so by linking whatever you're trying to remember with fear or excitement or some other emotion. What do I mean? Suppose you're trying to remember that you have to drop a letter in the mailbox first thing in the morning. You could tell yourself, *Don't forget to mail the letter!* But you'll be more likely to "not forget" if you instead tell yourself that if you don't mail the letter, you'll fall off a cliff when you step out your front door. Imagine the scene as vividly as you can. Chances are, as you leave your house, that image will pop to mind, and so will the memory of what you need to do.

Group 'Em

Remembering long lists is no easy task. But you'd be surprised at how well you can do if you group items into sets of no more than four, a tactic that's called "chunking."

If you're trying to remember the string of numbers 454699077777, you might think of it as 4546-9907-7777. Similarly, if you're trying to remember a grocery list, group your buys in some memorable way, like where they are in the grocery store. So, eggs, napkins, paper towels, chicken, cheese, steak, milk, fish, and toilet paper becomes much more memorable as eggs, milk, cheese; napkins, paper towels, toilet paper; chicken, steak, fish.

Make an Association

We've all heard the suggestion that in order to remember someone's name, we should associate it with a specific attribute. But it's not enough to simply remember Bob as blue-shirt Bob. What if you come across him and he's changed his attire? And what's to distinguish him from other people in blue shirts?

A more effective strategy is to choose a facial or other unchanging feature and create a memorable image in your mind that incorporates the person's name. For example, to remember Bob, you might note his rather short legs and imagine him bobbing up and down as he tries to reach something on a tall table. (If he's tall, you might have to pick some other attribute or alter this one in some memorable way.)

The same trick can apply to other information you're trying to remember. If you want to remember your cousin Harry's birthday in late October, you might try to think of him as Halloween Harry and picture him dressed as a goblin and carrying a candy bag. You can also tie someone's name to his or her profession, a trick I use when I meet people in social settings. Meeting Penny, a writer, I might think of a pen to remember her. Use your imagination. You'll be surprised at what you come up with.

Make an Effort

This may sound obvious, but we've all lamented our inability to remember a name when, in fact, we really didn't put much effort into giving it attention in the first place. Tell your hippocampus something is important—and worthy of memory—by repeating the information several times, writing it down, or in some other way making an effort to remember it. In the case of remembering someone's name, ask the person to repeat it or even to spell it. Then say it back to him or her. With each repetition you're sending a "save" signal to your brain.

Putting the Tips to Use: One Technique

In the late spring of 2012, I was invited to share the science behind brain fitness with the studio audience of cardiothoracic surgeon and television personality Dr. Mehmet Oz. This being TV, just having me *tell* the audience what I knew would be far too boring. Instead, I opted to show them. (You can watch this by going to my website, www.neurologyinstitute.com.)

Before the show, Dr. Oz's production team randomly selected seven audience members to learn a few tools they could use in practicing feats of memory. We met in a comfortable conference room and got right to work.

There are a slew of highly effective memory techniques that range from the simple to the extremely complex, but since our goal—to remember a list of twenty words—was fairly easy, I chose one of the simplest. First, I picked up a marker and wrote on a whiteboard twenty random words, selected by the production team. Then the fun began.

We first "chunked" the list into groups of four. Then, together, we imagined a funny scene for each chunk and placed it, in our minds, behind the chair of a fellow group member. (This is a variation of the memory palace technique, which I had to adapt since the group of people had just met and didn't have a common place they could all envision in their minds.) The first chunk—milk, egg, apple, hamburger meat—we placed behind the chair of Stacey, who was sitting to my right. Next, we placed corn, grapes, pineapple, and chicken behind the chair of Lisa. In short order, we ran through the list and placed images behind the chairs of each of the people sitting at the conference table. It only took a few repetitions before my group of memorizers had it down.

Then it was showtime. Dr. Oz bounded onto the stage and we launched into a lively discussion of brain health, including the benefits of "working out" your memory skills. I declared that memorizing long lists of random items isn't actually all that hard. That anyone could do it. Of course, it was time to prove my point. Dr. Oz called Stacey onto the stage, and without missing a beat Stacey recited the twenty words in perfect order. Then she did it in reverse order.

Once you've learned a technique, memorization is easy. And, as you now know, the more you do it, the better you get. Practicing is so simple it's almost child's play. (In fact, I've taught my own children simple memory techniques and encourage them to practice.)

Give it a try. Make a list of twenty random items, chunk them in groups of four, create evocative images, create a story for each chunk, and away you go.

Practice reciting the stories until you've committed them to memory. (Hint: It shouldn't take long!)

1. ______________________________

2. ______________________________

3. ______________________________

4. ______________________________

5. ______________________________

6. ______________________________

7. ______________________________

8. ______________________________

9. ______________________________

10. ______________________________

11. ______________________________

12. ______________________________

13. ______________________________

14. ______________________________

15. ______________________________

16. ______________________________

17. ______________________________

18. ______________________________

19. ______________________________

20. ______________________________

Brain-Building Activities

There are more brain-building activities than I could ever hope to list here. But I've included some examples of activities, grouped by level of difficulty. In general, the harder your brain works, the bigger the benefit, although keep in mind that even small efforts add up over time. I'd love to see you memorize a deck of cards, but if all you can manage for now is a puzzle, it's far better than throwing in the towel and doing nothing.

Easy	Moderate	Hard
Playing cards with friends Puzzles Easy sudoku Listening to podcasts that teach about a subject Sewing Cooking according to a recipe	Harder sudoku, crosswords, or online brain-teasing games Dancing Geo-caching,[12] letterboxing, or orienteering Learning a new sport Trip planning Volunteering Carpentry	Memorizing a deck of cards Learning a new language Memorizing all the states and their capitals or all the world's countries and their capitals Taking a college course (coursera.org, for example, offers free online courses from some of the most elite universities) Memorizing fifty random words in order (and backward)

Your Rx

Track 1

If you're a cognitive couch potato—you do little to stimulate your brain, aren't socially active, and rarely attempt to memorize facts or names—this track is for you.

Week One: Practice memorizing a list of eight random words a day, three days a week, using my tips above. (Hint: Use a newspaper or magazine to help you pick random words for memorization.)

Week Two: Practice remembering a list of twelve random words a day, three days a week.

Week Three: Practice remembering twenty random words a day, three days a week. Try to incorporate cognitive stimulation into your daily life as well. Think of it as the equivalent of taking the stairs rather than the elevator: add numbers in your head, play around with your electronic gadgets to see what new functions you can perform with them, read the directions and figure out how things work!

Track 2

If your life is somewhat socially and cognitively stimulating but you rarely take on activities that stretch your "memory muscle," this is your track.

Week One: Practice memorizing a list of twenty random words a day, four days a week. Try to incorporate cognitive stimulation into your daily life as well. See my suggestions in Week Three, above.

Week Two: Practice remembering four new names and faces a day, four days a week.

Week Three: Practice remembering five new names and faces a day, five days a week.

Track 3

If you're socially active, your life or work involves mental gymnastics, and you're already good at remembering names, this is your track. Your goal is to flex your "memory muscle" in new and challenging ways in order to enhance your cognitive flexibility.

Week One: You're already good at this, so practice remembering thirty-six items three days a week, and ten names two days a week.

Week Two: Pick a new hobby or take a new class. Spend one hour this week learning a new skill: maybe a new language, wine tasting, bird watching, photography, dancing—whatever you like. Remember, the best options are those that cross-train both your body and brain.

Week Three: Learn to memorize a deck of cards. Nelson Dellis explains his method online: http://climbformemory.com/2010/06/08/how-to-memorize-a-deck-of-cards, but other mnemonists describe different techniques online and in books. Pick one and start practicing!

PART III
Your Plan in Action

CHAPTER EIGHT

Ready, Set, Go

NOW THAT YOU know the science behind a bigger brain, you're ready to start building that eight-cylinder engine beneath your skull. Over the next twelve weeks, you'll implement a custom-crafted plan aimed at boosting BDNF, increasing oxygen flow, and training your brain to work within the ideal alpha zone.

You already got a glimpse of the plan, the track system that I previewed at the end of each chapter. Now you'll fully implement your own plan, following your selected track for each brain-boosting area each week and rating yourself on your effort. I recommend that each Sunday evening you rate yourself on the week before and plan your brain-building efforts for the week ahead.

By the time you reach week twelve, your weekly score will have risen, along with your Fotuhi Brain Fitness Score (which you will calculate at the end of this chapter and can compare to your baseline score from chapter 3), and you'll have the enhanced memory, clarity, and creativity to show for it.

First, a little more detail. In contrast to your Fotuhi Brain Fitness Score, which represents your current brain fitness based on your current lifestyle, you'll now be rating yourself on your *effort* each week using a scale of 1 to 5 (with 5 being the best) in eight key categories:

Exercise Give yourself a 1 if you were sedentary this week, or a 5 if you were quite active, engaging in thirty minutes of high-intensity aerobic exercise, combined with 15 minutes of weight lifting five days a week. Give yourself a 2, 3, or 4 if you fell somewhere in between.

Food Quality Give yourself a 1 if you ate a very unhealthy diet this week (no fruits or vegetables, no fish, and items high in trans fats, salt, and simple carbohydrates), or a 5 if your diet was healthy (balanced, with five daily servings of fruits and vegetables, two to three servings of fish a week, and items high in flavonoids and low in trans fats, salt, and simple carbohydrates). Give yourself a 2, 3, or 4 if you fell somewhere in between.

Food Quantity Give yourself a 1 if you ate oversize portions and paid no attention to the number of calories you consumed this week, or a 5 if you were cognizant of caloric intake and ate healthy portions (a serving of meat, for example, that was the size of the palm of your hand). Give yourself a 2, 3, or 4 if you fell somewhere in between.

DHA Give yourself a 1 if you ate no fish and did not take a DHA supplement this week, or a 5 if you ate fish twice this week and took 1,000 milligrams of DHA daily. Give yourself a 2, 3, 4 if you fell somewhere in between.

Mindfulness Give yourself a 1 if you were stressed every day this week, a 5 if you were often in the alpha zone (focused, calm, and alert), or a 2, 3, or 4 if your state of mind fell somewhere in between.

Attitude Give yourself a 1 if you were frequently pessimistic this week, a 5 if you were optimistic, or a 2, 3, 4 if you fell somewhere in between.

Sense of Curiosity Give yourself a 1 if you did little to expand your horizons, learn something, or try a new activity; a 5 if you were highly curious, learning, and trying new activities; or a 2, 3, 4 if you fell somewhere in between.

Memory Stimulation Give yourself a 1 if you didn't make any effort this week to memorize names, phone numbers, lists, or anything else; a 5 if you memorized the name of every person you met and always memorized your shopping list or other random lists; or a 2, 3, or 4 if your memory training fell somewhere in between.

In addition to your scores, you should note your blood pressure and pulse each week.

You'll also use the "notes" section of the scorecard to record any unusual factors. If a work project caused you heightened stress one week, an injury affected your mobility, or a new medication affected your sleep, this is the place to register it! You can also use this space to keep track of your weight or clothing size.

When you rate yourself each week, you'll obtain a total score. To track your progress over the next twelve weeks, plot your weekly total score on the graph on the next page.

Week One

It usually takes about three weeks to form a habit; hence, the first three weeks of the plan outlined next are fairly concrete, while the remaining weeks offer some flexibility. For the first week, you'll complete your chosen track in each

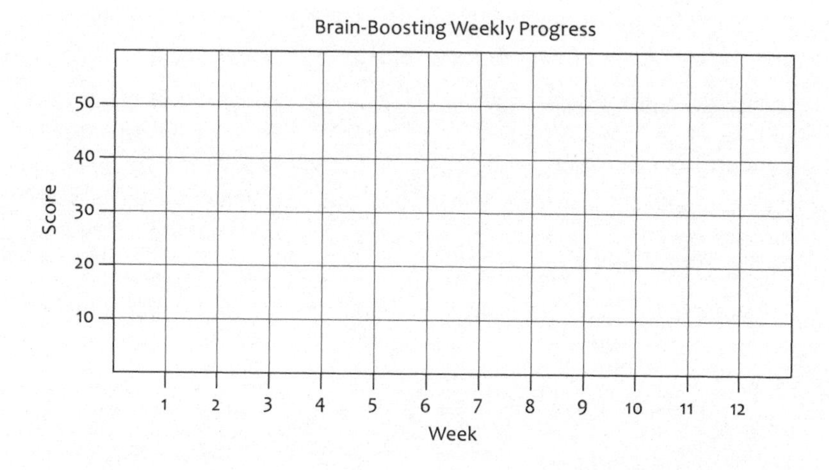

of the four brain-booster categories. As the weeks progress, each track will get progressively more ambitious. If at any point you feel you could do more, move up to the next track and complete the assigned work for that track.

As you work through the weeks, continue to read this book to learn about brain shrinkers and see if they apply to you. I've noted a few weeks where you should pay particular attention to a few common brain shrinkers.

Now choose your track for each category and follow the advice. You've already seen a preview of these, but we've reprinted them here to make selection easier.

Exercise

Track 1: If you lead a sedentary lifestyle and are currently not exercising at all, this is your track. I want you to get moving, but I don't expect you to run a marathon (at least not anytime soon). Start by walking for ten minutes at a time, three days a week.

Track 2: If you're not completely sedentary—for example, if you walk a good bit for work—and exercise occasionally, this is your track. Walk fast or jog for twenty minutes, three days a week. (Don't count time you walk at work; this must be twenty uninterrupted minutes.)

Track 3: If you already exercise twice a week or more, this if your track. Do thirty minutes of uninterrupted aerobic activity, five days a week, plus five extra minutes of weight lifting, push-ups, or another muscle-building activity, three days a week.

Diet

Track 1: If you're currently consuming a highly unhealthy diet, with few brain-building nutrients, this is your track. First, you'll need to detox,

Exercise Tips

MAKE IT AN EVENT Even if you're on the easy end of track 1, take your exercise seriously. Put on comfortable shoes, schedule time, and then get outside, on a treadmill, in a mall, or wherever you choose to walk or run. Creating habits—like setting aside time to move and even breaking out and lacing up your sneakers—that center around an activity will make you much more likely to stick with it in the long run.

SUBSTITUTE Cross-train by substituting stationary cycling, swimming, or another aerobic activity for walking or running.

BUDDY UP Find an exercise partner. You're less likely to cancel a workout or quit your regimen if someone else is counting on you to show up. Set goals together and celebrate when you reach them.

cutting the worst foods from your diet before you make any serious effort to add brain builders. Commit to cutting your consumption of trans fats almost entirely and your consumption of simple carbohydrates—often found in processed foods—by 50 percent. Try to avoid sugary foods, such as donuts, which are high on the glycemic index, as they'll cause a spike in your blood sugar.

Track 2: If you've already detoxed, or you already eat a somewhat healthy diet, this is your track. Focus on eating natural rather than processed or fast food. Be sure to include flavonoids.

Track 3: If you're already eating a healthy, balanced diet, this is your track. Supplement your diet with 1,000 milligrams daily of DHA. (Note: If you take Coumadin or another blood thinner, do not take DHA.) Be sure you know the levels of your vitamins B12 and D. If they're low, supplement. If they're borderline, adjust your diet to feature more foods that offer vitamins B12 and D.

Mindfulness

Track 1: If you've never done any type of mindfulness activity, this is your track. Start with my simple 7-7-7 breathing exercise. Do it once a day, three times a week, and focus on your technique. If you feel noticeably calmer after doing it, you've probably got it right.

Track 2: If you've had some experience in the past with meditation or another form of mindfulness, or are currently taking such a class, this is your track. Meditate twenty minutes a day, four days a week.

Track 3: If you're already a practiced meditator or regularly engage in some other mindful activity, this track is for you. Continue with your mindfulness practices and set aside three half-hour sessions to write down all the things you love about your life. Think about why you love them and then come up with ways to increase your exposure to the things you cherish. For example, I love spending time with my daughters, Nora and Maya, so one of my thirty-minute sessions might end with a plan that allows me to get home from work a bit earlier twice a week.

Cognitive Stimulation

Track 1: If you're a cognitive couch potato—you do little to stimulate your brain, aren't socially active, and rarely attempt to memorize facts or names—this track is for you. This week, practice memorizing a list of eight random words a day, three days a week, using the tips provided in chapter 7. (Hint: Use a newspaper or magazine to help you pick random words for memorization.)

Track 2: If your life is somewhat socially and cognitively stimulating but you rarely take on activities that stretch your "memory muscle," this is your track. This week, practice memorizing a list of twenty random words a day, four days a week. Try to incorporate cognitive stimulation into your daily life as well. Think of it as the equivalent of taking the stairs rather than the elevator: add numbers in your head, play around with your electronic gadgets to see what new functions you can perform with them, read the directions and figure out how things work!

Track 3: If you're socially active, your life or work involves mental gymnastics, and you're already good at remembering names, this is your track. Your goal is to flex your "memory muscle" in new and challenging ways in order to enhance your cognitive flexibility. This week, practice remembering thirty-six items three days a week, and ten names two days a week.

Rate Yourself

Week 1

Date ____________

At the end of week one, please rate your efforts in the following areas.

						Notes:
Exercise	1	2	3	4	5	
Food Quality	1	2	3	4	5	____________
Food Quantity	1	2	3	4	5	____________
DHA	1	2	3	4	5	____________
Mindfulness	1	2	3	4	5	____________
Attitude	1	2	3	4	5	____________
Sense of Curiosity	1	2	3	4	5	____________
Memory Stimulation	1	2	3	4	5	____________
TOTAL SCORE:	____________					____________
Blood pressure:	____________					____________
Pulse:	____________					____________

Act Your Age!

As we age, different factors affect our brain health and performance. As you continue through your twelve-week plan, keep these age-based modifications in mind:

In Your Twenties and Thirties

Most people in this age group have minimal memory and cognitive function complaints, although attention can be a concern. The chief goal now is to get the brain into the best possible condition to ensure maximum performance now and to offset the wave of aging that's ahead. At this age you should . . .

Work on Attention Issues Regular exercise and daily supplementation of 1,000 mg of DHA may help reduce problems with attention. But if such interventions don't seem to help, be sure to talk to your doctor about your attention concerns. You may have treatable conditions such as sleep apnea, an underactive thyroid, vitamin deficiency, or ADHD. These could be treated well with simple interventions that may or may not include medication.

Build Brain-Healthy Lifetime Habits They'll benefit you now and in the future, so the sooner you make it a lifestyle, the better. Just like you have a routine of brushing your teeth, get into the habit of eating brain-healthy food, exercising regularly, watching your waist-to-height ratio, getting quality sleep, and stimulating your brain.

Get Emotionally Fit Find strategies that work for you to reduce stress. Set long-term goals you would love to achieve one day, and then take pleasure in completing small steps toward those goals on a daily basis. If your long-term goal is to be a lawyer, for example, take pleasure in the fact that you study an extra hour for a test in the basic college course that will help you further your academic career.

Treat Any Brain-Draining Conditions If you have or suspect you have sleep apnea, diabetes or pre-diabetes, hypertension, high cholesterol, or any other vascular risk factors, be diligent about treating them and reducing your risk. The effects of these diseases are cumulative, so the longer you have them the more you've damaged your brain.

In Your Forties and Fifties

Women in their late forties to midfifties can expect to go through menopause, which can cause temporary brain fog and memory and attention lapses. Men can also suffer from low testosterone, which can affect memory. (A simple test can detect low testosterone, a treatable condition.) This is the stage of life during which the normal aging process begins to chip away at brain size, often without any obvious symptoms other than lapses in memory (like the keys you lost last week). Therefore, at this age you should . . .

Stay Physically Fit It's not too late to run a marathon or train for a triathlon, but extreme fitness efforts aren't necessary. Join a walking group, enroll in a dance class, or learn how to step up your tennis game.

Stay Emotionally Fit This is a good time to check the health of your relationships—with your spouse, your coworkers, your friends, and your children. Ensuring they're stable and rewarding can help you stay in the alpha zone. Stress reduction is as important as ever. Try yoga, meditation, or calm breathing exercises.

Sleep As we age, sleep problems become more common. Often we set up conditions that make sleep deprivation a way of life. If you're working until midnight every night and getting up at five in the morning, you may find that you routinely short yourself on sleep. Be sure you're getting seven to eight hours of quality sleep every night.

In Your Sixties and Beyond

With age, your risk of developing many chronic health problems increases. Not only that, but your brain also encounters more of the effects of normal aging. To offset that reality . . .

Stay Physically Fit Sixty-, seventy-, and even eighty-year-olds continue to astound their peers with what they can do in sports. But some slowing down physically is a natural part of aging for many people. Find an exercise—walking, water aerobics, dance—that still works for you. And work it!

Stay Healthy Be sure to monitor your health and work with your doctor to aggressively treat conditions such as hypertension, diabetes, high blood cholesterol, and heart disease. The elderly also may have trouble absorbing certain nutrients, so be sure you're on top of health measures for factors that can affect cognitive function (vitamins B12 and D and thyroid function are a few).

Take DHA This is important at any age but particularly so as we age.

Stay Socially Engaged Eighty is the new sixty. Elderly people today find themselves drawn more than ever to participate in social engagements; they don't have the burden of child-rearing and have the physical and financial capacity to be highly active. Join this new movement and sign up for group travel, cooking classes, social networking, and more.

Week Two

Exercise

Track 1: Increase your walking time to fifteen minutes, three times a week.

Track 2: Increase your walking or jogging time to thirty minutes, four days a week. If you're walking, try to jog at least some of the time. Increase the amount of time you jog versus walk over the next ten weeks.

Track 3: Continue with thirty minutes of uninterrupted aerobic activity, five days a week. Bump up your muscle building to ten minutes, three days a week.

Diet

Track 1: Continue to stay away from junk food, and now cut down salt and cholesterol. Choose lean meat and eat it no more than once or twice a week.

Track 2: If you drink coffee, cut back to one cup in the morning and none in the afternoon. Instead, add a cup of tea in the afternoon.

Track 3: Continue with DHA supplementation. Don't forget to get adequate water. Your brain needs water to function well, so be sure to drink six to eight glasses a day. Add a glass of wine per day several days a week.

Mindfulness

Track 1: Continue with my 7–7–7 breathing exercise and add one session of meditation.

Track 2: Meditate twenty minutes a day, four days a week, and add a tai chi or yoga class once a week.

Track 3: Continue with your mindfulness practices and set aside three half-hour stretches to write down the things that bother you. Fill out an ABC chart for each and challenge your assumptions about your belief system. Ask yourself, *Why am I unhappy? How can I change my perception?* Come up with strategies to change or decrease your exposure to the things that make you unhappy.

Cognitive Stimulation

Track 1: Practice remembering a list of twelve random words a day, three days a week.

Track 2: Practice remembering four names and faces a day, four days a week.

Diet Tips

TRACK IT Most adults need between 2,000 and 2,500 calories a day. Consider tracking what you eat each day for one week, using a food journal or one of the many available phone apps or online tools. Tracking what you eat is helpful in two ways. First, it forces you to be more aware of what you're putting in your body. Second, tracking your food consumption can help you identify patterns in your eating (too few leafy greens?) that you need to address.

CUT THE COFFEE If you have a serious coffee habit, slowly reduce your coffee consumption until you're drinking just one cup a day. Add one cup of herbal tea in the afternoon as well.

STAY HYDRATED Your body—your brain included—needs water to function well, so make an effort to consume six to eight glasses of water daily.

AVOID HIGH-CARB DIETS A high-carbohydrate diet can cause a spike in your blood glucose level, which damages blood vessels and neurons and has been associated with inflammation and weight gain.

Track 3: Pick a new hobby or take a new class. Spend one hour this week learning a new skill: maybe a new language, wine tasting, bird watching, photography, dancing—whatever you like. Remember, the best options are those that cross-train both your body and brain.

Rate Yourself

Week 2

Date ________________

At the end of week two, please rate your efforts in the following areas.

		Notes:
Exercise	1 2 3 4 5	
Food Quality	1 2 3 4 5	________________
Food Quantity	1 2 3 4 5	________________
DHA	1 2 3 4 5	________________
Mindfulness	1 2 3 4 5	________________
Attitude	1 2 3 4 5	________________
Sense of Curiosity	1 2 3 4 5	________________
Memory Stimulation	1 2 3 4 5	________________
TOTAL SCORE:	____________	________________
Blood pressure:	____________	________________
Pulse:	____________	________________

Week Three

You're almost into habit territory. Keep up the good work, increasing your effort in all brain-boosting areas.

Exercise

Track 1: Increase your walking time to twenty minutes, three days a week. In addition to your scheduled time, find ways to work activity into every day. Take the stairs, park at the far end of the parking lot, walk to lunch, go for a hike instead of going to the movies–do anything you can to replace sedentary moments with activity.

Track 2: Increase your walking or jogging time to thirty minutes, five days a week. Gradually increase the intensity of your workouts. Ideally, you want to get to a level of intensity to achieve 60 to 80 percent of your maximum heart rate. If you'd like, you can substitute a stationary bike, swimming, or another aerobic activity, or you can substitute a one-hour game of tennis (or another intense activity) for two of your thirty-minute sessions.

Track 3: Continue with thirty minutes of uninterrupted aerobic activity, five days a week. Bump up your weight lifting to fifteen minutes, five days a week. Increase the intensity of your workouts as needed so that you're improving your physical fitness. Consider adding high-intensity interval training (HIIT), consisting of five bursts of vigorous exercise during the last fifteen minutes of your exercise routine (five bursts for one minute each, followed by two minutes of less intense activity). You should feel tired after you finish your exercise.

Diet

Track 1: Continue with the cuts you made in weeks one and two and work on reducing your serving sizes and limiting caloric intake. If you haven't already, try tracking your calorie intake for one week through an online tool or app. Seeing just how much a bagel and cream cheese "costs" you in calories can help you convince yourself to tuck into a bowl of yogurt with blueberries instead.

Track 2: Focus on adding salmon or other fish to your diet, and add more fruits and vegetables.

Track 3: Continue with DHA supplementation, and continue to eat healthy portions and brain-friendly foods. Try new flavonoids you've never tried before.

Mindfulness

Track 1: Shift from breathing exercises to twenty minutes of meditation, four days a week.

Track 2: Continue with meditation and add to your week one more session of tai chi, yoga, or another relaxing activity.

Track 3: Give yourself a Sabbath. From sunset one day until sunset the following day, promise to do no work. This is your fun time, so fill it with things that make you happy.

Cognitive Stimulation

Track 1: Practice remembering twenty random words a day, three days a week. Try to incorporate cognitive stimulation into your daily life as well. Think of it as the equivalent of taking the stairs rather than the elevator: add numbers in your head, play around with your electronic gadgets to see what new functions you can perform with them, read the directions and figure out how things work!

Track 2: Practice remembering five new names and faces a day, five days a week.

Track 3: Learn to memorize a deck of cards. Nelson Dellis explains his method online (http://climbformemory.com), but other mnemonists describe different techniques online and in books. Pick one and start practicing!

Mindfulness Tips

LEARN A TECHNIQUE Seek out a meditation, yoga, tai chi, or drumming class locally, or through online instruction, or a CD or DVD.

FIND A SPACE Find a quiet, comfortable practice place with few distractions. You can also try walking meditation, which you can do in a garden or quiet neighborhood.

OPEN YOUR MIND For meditation to work, you'll need to have an open mind and the ability to let go of any distractions or negative thoughts as you meditate. Practice this; it becomes easier over time.

LIVE IT Once you've learned meditation techniques, use them to shift yourself into the alpha zone throughout your day. Your goal is to frequently feel focused, calm, and alert. Smile as often as you can.

DO YOUR ABCs Schedule time to do the ABC technique I described in chapter 6. Think about how you can challenge your beliefs and reduce anxiety; it's your ticket to the alpha zone.

Rate Yourself Week 3

Date ______________

At the end of week three, please rate your efforts in the following areas.

		Notes:
Exercise	1 2 3 4 5	
Food Quality	1 2 3 4 5	______________
Food Quantity	1 2 3 4 5	______________
DHA	1 2 3 4 5	______________
Mindfulness	1 2 3 4 5	______________
Attitude	1 2 3 4 5	______________
Sense of Curiosity	1 2 3 4 5	______________
Memory Stimulation	1 2 3 4 5	______________
TOTAL SCORE:	______________	______________
Blood pressure:	______________	______________
Pulse:	______________	______________

Week Four

Congratulations, you've completed the first three weeks of the program, building habits—and your noggin—as you go.

If you ended week three on track 1 or 2, move up to the next track this week. If you ended it on track 3, simply repeat week three of track 3 for the remainder of the program, adding intensity and variety whenever possible.

Rate Yourself Week 4

Date ______________

At the end of week four, please rate your efforts in the following areas.

		Notes:
Exercise	1 2 3 4 5	
Food Quality	1 2 3 4 5	______________
Food Quantity	1 2 3 4 5	______________
DHA	1 2 3 4 5	______________
Mindfulness	1 2 3 4 5	______________
Attitude	1 2 3 4 5	______________
Sense of Curiosity	1 2 3 4 5	______________
Memory Stimulation	1 2 3 4 5	______________
TOTAL SCORE:	______________	______________
Blood pressure:	______________	______________
Pulse:	______________	______________

Cognitive Stimulation Tips

CONSIDER YOUR CHOICES Spend some time thinking about your likes and dislikes. Then make a list of five activities you really enjoy and consider how well each will stimulate your brain. Choose the ones that are highest in appeal *and* brain-building potential.

MAKE IT A HIGH-PRIORITY HABIT It's important to train your "memory muscle," so take advantage of every opportunity to memorize new information and learn new things.

KEEP IT FRESH Once you've mastered a skill or task, move on or find a way to introduce a new challenge. If you've played golf every week for a decade, continue to enjoy it, but also try bowling or another new activity. Have you gotten pretty good at playing the piano? Try tackling a new piece that's above your skill level. As a general rule, if you're slightly out of your comfort zone when you tackle a new task, you're challenging your brain—in a good way—and making the new synapses that will add to your brain reserve.

LIMIT DISTRACTIONS Living in the wired world means we're constantly interrupted by our devices. Often we make things worse by initiating our own interruptions—checking e-mail or logging into Facebook obsessively, for example. Schedule set times—once an hour, perhaps—to check e-mail, respond to your electronic devices, or enjoy online social networking. Resist the temptation to return to your old, obsessive ways.

EXPLORE YOUR WORLD When you bring a child into a new room, she'll set to work immediately, picking up items, examining them, putting them down. She'll peek and poke and prod at everything in sight. As adults, experience tells us there's nothing of interest behind the curtains, but while we don't need to peer in every corner, maintaining curiosity about the world can help us keep our brains engaged and growing. So, explore your world. If you can't travel, get online and take a virtual tour of a city half a world away. Read *National Geographic* magazine. Download a new app and figure out how it works. Be curious.

BECOME A BETTER SOCIALIZER If you find it difficult to make conversation or uncomfortable to be in social situations, give yourself tools that make it easier. I have a shy friend who comes to every party with a new joke to tell. He always gets a laugh, which makes him feel good. If you're not a jokester, try boning up on a favorite topic so you can comfortably add to the conversation.

Checkup Time!

Good work on completing the first month of your bigger-brain plan. This is an ideal time to spend a few minutes thinking about what has worked so far—and what hasn't. Jot down three major accomplishments you're proud of and three challenges you still need to tackle.

Accomplishments

1. ______________________________
2. ______________________________
3. ______________________________

Still to do

1. ______________________________
2. ______________________________
3. ______________________________

Week Five

Now that you've created some habits around brain boosters, it's time to turn your attention to avoiding brain drainers.

You'll read about them starting in Part IV, but for this week let's focus on sleep quality. One common condition that robs people of quality sleep is obstructive sleep apnea (OSA), a sleep disorder that literally starves the brain of oxygen and shrinks the hippocampus. Most alarming of all is that many people have OSA and don't realize it.

You'll learn more about OSA in chapter 9, but for now you should know that risk factors for OSA include being overweight, having a neck circumference of more than seventeen inches in men and sixteen inches in women, smoking, alcohol abuse, and structural issues, such as a large tongue or an overbite. Other signs that you might suffer from OSA include daytime sleepiness, waking yourself with your own snoring, being told by someone else that you snore loudly (especially if it's a snore that ends in a gasp for breath), difficulty concentrating, depression, or irritability.

If you think you have OSA, be sure to seek treatment immediately, as this brain shrinker may be offsetting the gains you've earned through your brain-boosting efforts.

Sleep Tips, Part One

If you have sleep apnea, seek treatment from a medical professional, but also follow these tips:

LOSE WEIGHT Although there are other causes of sleep apnea besides being overweight, weight loss can be highly effective in reducing obstructive sleep apnea.

AVOID ALCOHOL IN THE EVENING Alcohol can relax the muscles in your upper airway, thus worsening obstructive sleep apnea. Avoid consuming alcohol in the hours before you go to sleep. Sedatives like those found in sleep aids can also pose problems for people with sleep apnea and should be avoided.

USE A CPAP Continuous positive airway pressure (CPAP) devices blow air into your nose while you sleep. If you have a diagnosis of sleep apnea, your doctor will prescribe a CPAP for you.

SLEEP ON YOUR SIDE Side sleeping can help reduce sleep apnea, as doing so keeps your tongue from falling backward and blocking your airflow. Make a pocket on the inside of the back of an old shirt and fill it with a few tennis balls—they'll prevent you from rolling onto your back while you sleep.

STOP SMOKING Smoking can increase inflammation in the upper airway, increasing your risk for obstructive sleep apnea.

Rate Yourself

Week 5

Date ____________

At the end of week five, please rate your efforts in the following areas.

		Notes:
Exercise	1 2 3 4 5	
Food Quality	1 2 3 4 5	____________
Food Quantity	1 2 3 4 5	____________
DHA	1 2 3 4 5	____________
Mindfulness	1 2 3 4 5	____________
Attitude	1 2 3 4 5	____________
Sense of Curiosity	1 2 3 4 5	____________
Memory Stimulation	1 2 3 4 5	____________
TOTAL SCORE:	____________	____________
Blood pressure:	____________	____________
Pulse:	____________	____________

Week Six

Just as OSA can wreak havoc with the quality of your sleep, so too can insomnia, which can also greatly reduce the quantity of your sleep. Signs of insomnia include difficulty falling asleep, a tendency to awaken extremely early, daytime sleepiness, moodiness and irritability, and difficulties with attention, memory, and concentration.

Insomnia can have many causes, including anxiety, depression, medication side effects, sleep disorders, excessive consumption of caffeine, chronic pain, or a host of other factors. If you wake up not feeling refreshed, wake up frequently during the night, or have trouble falling asleep, talk to your doctor about insomnia. This is a treatable condition.

Rate Yourself

Week 6

Date ______________

At the end of week six, please rate your efforts in the following areas.

		Notes:
Exercise	1 2 3 4 5	
Food Quality	1 2 3 4 5	______________
Food Quantity	1 2 3 4 5	______________
DHA	1 2 3 4 5	______________
Mindfulness	1 2 3 4 5	______________
Attitude	1 2 3 4 5	______________
Sense of Curiosity	1 2 3 4 5	______________
Memory Stimulation	1 2 3 4 5	______________
TOTAL SCORE:	______________	______________
Blood pressure:	______________	______________
Pulse:	______________	______________

Sleep Tips, Part Two

EXERCISE Exercising for a minimum of 150 minutes a week may help you sleep better and feel more alert during the day.[1]

AVOID LONG NAPS If you have insomnia, avoid taking naps. Daytime sleeping will make it difficult to fall asleep or stay asleep at night.

LIMIT USE OF ELECTRONICS IN THE HOUR BEFORE SLEEP TV and other electronics can rev up your brain before sleep, and the glow from such devices can actually reduce your production of melatonin, a hormone needed for slumber.

USE THE BEDROOM FOR SLEEPING ONLY Make it your sleep sanctuary. That means removing the TV and avoiding eating, working out, or working in your bedroom.

MONITOR YOUR DIET FOR THINGS THAT MAY BE KEEPING YOU UP Too much caffeine during the day, eating spicy food, or having a large meal right before bedtime are possibilities to consider and remedy. Some medicines can also cause insomnia.

AWAKEN AND GO TO SLEEP AT ROUGHLY THE SAME TIME EVERY DAY Waking up at the same time, in particular, is important; so if you experience insomnia or a night of insufficient sleep, try to rise at your usual time and go to bed earlier the next night. If your schedule isn't permitting you a solid seven to eight hours sleep on a regular basis, make the changes needed to ensure that it can.

CHILL OUT A too-hot room can keep you awake, so try to sleep in a room that is around 65°F. Taking a hot shower three or four hours before you sleep can also help to lower your body temperature and relax your muscles.

DON'T STAY IN BED AWAKE If you're having trouble falling asleep, don't lie in bed watching the clock. Instead, get up and engage in a quiet activity like reading. Then go back to bed when you feel sleepy.

SLEEP UNTIL SUNLIGHT If you can, wake up with the sun, or use very bright lights in the morning to help reset your body's internal clock.

Week Seven

You've now completed your first six weeks of the brain fitness program. By this point, you have made progress—and you're feeling the brain benefit. You no longer worry about your memory, as it has improved significantly. But remember, you're only halfway there; more progress is ahead.

Keep adding intensity or expanding on your efforts as much as possible and continue your progression up the tracks.

Rate Yourself

Week 7

Date ______________

At the end of week seven, please rate your efforts in the following areas.

		Notes:
Exercise	1 2 3 4 5	
Food Quality	1 2 3 4 5	______________
Food Quantity	1 2 3 4 5	______________
DHA	1 2 3 4 5	______________
Mindfulness	1 2 3 4 5	______________
Attitude	1 2 3 4 5	______________
Sense of Curiosity	1 2 3 4 5	______________
Memory Stimulation	1 2 3 4 5	______________
TOTAL SCORE:	______________	______________
Blood pressure:	______________	______________
Pulse:	______________	______________

Week Eight

As you'll soon read, excess stress is a major brain shrinker, leading to reduced cognitive function in both the short and long term.

This week, focus on reducing your stress level (if you have excess stress). One way to do that is by identifying your stressors and finding ways to reduce them. My favorite stress-reduction tool is the stress chart. Set aside an hour and create your own stress chart by following the steps I've included in the next table.

Create a list with a column for "buckets," or categories for the different areas of your life. Your buckets might be family, work, finances, extended family, household, or health.

For each bucket item, identify the top five to ten things that cause you stress. You can call this column "What's stressful."

Create a column for your goals. (Call this one "My goal.") Then go through each item and come up with a realistic plan for how you can reduce your exposure to the stressor. Keep in mind that, in some cases, a perfectly reasonable solution may be to decide not to get involved.

Your list might look like this:

Stress Chart

Bucket	What's stressful	My goal
Family	Finding time to spend with the kids	Schedule in two "Mommy time" or "Daddy time" events per week. (Consider baking, crafting, sports, board games, or other activities.)
	Finding time to be alone with my spouse	Budget for a babysitter. Set aside two nights a month for a date.
	Arguing with my spouse about child care and housework	Write down all household responsibilities; make an agreement on division of duties.
	Trying to get the kids to do their homework	Declare quiet homework time from 4:00 to 5:00 P.M. daily. Turn off all electronics and enforce homework time.
Work	Trying to leave every night by 5:30 P.M. and still get my work done	Pick two nights a week to stay late; leave at 5:30 P.M. the other nights, no exceptions. Ask the boss about shifting my schedule; make this a priority.
	Horrible commute	Ask about telecommuting. Use the time for brain-boosting activities, such as learning a new language or listening to prerecorded lectures.
	Extra projects	Talk to the boss about my workload. How can we better plan to avoid "extra" projects?
Finances	Never enough money!	Accept that this will always be true for our family. Research better budgeting, make a budget, and stick to it.
Household	Trying to get the kids to help me around the house	Set up a reward system to encourage help. Stop doing everything for them.
	Laundry overflowing	Consider hiring a mother's helper to watch the baby two times per week while I do laundry. Or budget to have someone do the laundry for us.
	Broken things around house	Make a list of the top four; take two and give two to my spouse to do by the month's end.

Bucket	What's stressful	My goal
Health	Worried about my weight	See a doctor to talk about it. Join a gym; make this a priority!
	Worried about my diet	Research nutritional meals. Set healthy eating goals. Institute a reward system for staying on track.
Extended family and friends	Planning sister-in-law's baby shower	Delegate some jobs (to Mom? Other siblings?).
	Getting dragged into fights between my mom and my brother	Tell them I will no longer mediate.
	Pressured to go to my book club even though I don't have time to read	Bow out gracefully, or ask the book club members if I can "audit" the book club and show up even if I haven't read, or tell them I need to take a six-month leave.

Rate Yourself Week 8

Date ______________

At the end of week eight, please rate your efforts in the following areas.

		Notes:
Exercise	1 2 3 4 5	
Food Quality	1 2 3 4 5	______________
Food Quantity	1 2 3 4 5	______________
DHA	1 2 3 4 5	______________
Mindfulness	1 2 3 4 5	______________
Attitude	1 2 3 4 5	______________
Sense of Curiosity	1 2 3 4 5	______________
Memory Stimulation	1 2 3 4 5	______________
TOTAL SCORE:	______________	______________
Blood pressure:	______________	______________
Pulse:	______________	______________

Stress Reduction Tips

PLAN AHEAD Invest in a good phone app—or an old-fashioned paper planner—and schedule your days and weeks in advance. I usually set time aside the first day of each year and set ten goals I'd like to achieve that year. On the first Sunday of each month, I set goals I'd like to accomplish that month, and each week I spend fifteen minutes on Sunday nights jotting down what I'd like to accomplish the following week. Every morning I spend five minutes writing down tasks to complete that day that will move me toward my weekly, monthly, and annual goals. Knowing what to expect can greatly reduce your level of stress and help you avoid overscheduling.

LEARN TO SAY NO It's hard, especially if you've gotten yourself into the habit of helping out whenever asked, but learning how to say no to requests for your time or energy is key to reducing your stress.

LIMIT EXPOSURE TO PEOPLE OR SITUATIONS THAT CAUSE YOU STRESS Your stress chart will help, but ultimately you have to make a commitment to avoiding some people or situations that cause you stress.

CHANGE YOUR ATTITUDE When stressful situations present themselves, try to put them in perspective in the grand scheme of things. Accept that some things in your life are beyond your control. Do the ABC exercise and work toward changing your attitude and emotions. And practice being upbeat and positive.

DO SOMETHING ABOUT IT If you're concerned about famine in Africa, or that you pay too much in taxes, or that your child's school doesn't devote enough time to art, do something about it. It might be as simple as writing a letter or attending the next school board meeting to voice your opinion, but taking action can help you break a pattern of anxious thinking. This gives you the pleasure of being proactive, as opposed to the agony of rehashing the same negative thoughts.

PRACTICE INTROSPECTION It's easy to get caught up in our hectic lives, running from one commitment to the next without giving much thought as to why. Take some time every few weeks for a little introspection. Ask yourself, *What am I doing in life? Where am I going? What's working for me and what's not working? What is important to me and how can I achieve it?*

GET HELP IF YOUR STRESS SEEMS UNMANAGEABLE If your stress level seems out of proportion, or you're concerned that you might be depressed or suffering from an anxiety disorder, ask your doctor for a screening.

Checkup Time

Congratulations on completing week eight of your bigger-brain plan. By now, parts of your brain have rejuvenated and you feel sharper and more mentally nimble already. You have one month to go, so spend a few minutes thinking about what has worked so far—and what hasn't. Jot down three major accomplishments you're proud of and three challenges you still need to tackle.

Accomplishments

1. ______________________________

2. ______________________________

3. ______________________________

Still to do

1. ______________________________

2. ______________________________

3. ______________________________

Week Nine

We don't often think about the possibility of a stroke until late in life, but as you'll read in chapter 12, the conditions that lead to a stroke often develop long before any symptoms appear. No matter what your age, you should be aware of the factors that contribute to strokes and take steps to limit your risk.

Remember, a stroke is only the tip of the iceberg: although some blood vessels were blocked enough to cause it, many other blood vessels are likely half blocked and are at risk of becoming fully blocked in the future. Risk factors include hypertension, diabetes, carotid artery and peripheral artery disease, atrial fibrillation, heart disease, sickle cell anemia, high cholesterol, inactivity, obesity, and smoking. Heredity may also play a role.

Stroke Tips

AVOID HYPERTENSION High blood pressure is the biggest risk factor for stroke, so knowing your blood pressure reading, and keeping it regularly within the normal range, is critical.

BEWARE OF DIABETES MELLITUS (TYPE 2) Diabetes often co-exists with other risk factors, such as high blood pressure or obesity, but it's a brain shrinker on its own as well. If you are overweight, get checked for diabetes.

KEEP YOUR BLOOD FLOWING High cholesterol can clog arteries throughout the body, slowing the flow of oxygen to the brain. Be aware of your cholesterol levels, and strive to keep your HDL and LDL levels in the healthy range.

STAY HEART-HEALTHY Coronary heart disease, heart attack, and other cardiovascular disease substantially increase the risk of stroke. Inactivity, obesity, high blood pressure, and high cholesterol all increase the risk of both heart attacks and stroke.

QUIT SMOKING Smoking kills. And it damages the cardiovascular system, and hence increases the risk of heart attacks and stroke.

Rate Yourself

Week 9

Date ________________

At the end of week nine, please rate your efforts in the following areas.

		Notes:
Exercise	1 2 3 4 5	
Food Quality	1 2 3 4 5	________________
Food Quantity	1 2 3 4 5	________________
DHA	1 2 3 4 5	________________
Mindfulness	1 2 3 4 5	________________
Attitude	1 2 3 4 5	________________
Sense of Curiosity	1 2 3 4 5	________________
Memory Stimulation	1 2 3 4 5	________________
TOTAL SCORE:	____________	________________
Blood pressure:	____________	________________
Pulse:	____________	________________

Week Ten

Congratulations on completing nine weeks of the brain fitness program. You're three-quarters of the way to your goal. By now, you feel sharper—and you're not the only one who's noticed. Friends and family are complimenting you on your memory and clear thinking. You've added synapses, new neurons and blood vessels, and bolstered your brain highways. Your CogniCity is flourishing. If you ended week nine on track 2, try moving up to track 3. If you ended week nine on track 3, repeat the third week of track 3 for the remainder of the program, adding intensity and variety when possible.

Rate Yourself

Week 10

Date ________________

At the end of week ten, please rate your efforts in the following areas.

		Notes:
Exercise	1 2 3 4 5	
Food Quality	1 2 3 4 5	________________
Food Quantity	1 2 3 4 5	________________
DHA	1 2 3 4 5	________________
Mindfulness	1 2 3 4 5	________________
Attitude	1 2 3 4 5	________________
Sense of Curiosity	1 2 3 4 5	________________
Memory Stimulation	1 2 3 4 5	________________
TOTAL SCORE:	____________	________________
Blood pressure:	____________	________________
Pulse:	____________	________________

Week Eleven

You're approaching the end of your twelve-week plan. Now is a good time to revisit the healthy-body discussion we had in chapter 3. If you're overweight or obese—and haven't already taken steps to address your weight—work with your doctor to create a plan for the months ahead.

You should also look back over your pulse and blood pressure notes for the past ten weeks. Are they within the healthy range? If not, talk to your doctor about addressing any underlying health problems. And take a moment to consider any medications you're taking. Are they still necessary? If you're unsure, talk to your doctor.

And one last bit of advice as you head into your final weeks: Tell your friends and family about your brain-boosting efforts and successes. Encourage them to join you and start their own brain fitness programs. As you move out of the program, you'll find having friends and family living a brain-healthy lifestyle will help you to maintain the good habits you've established so far.

Rate Yourself

Week 11

Date ______________

At the end of week eleven, please rate your efforts in the following areas.

		Notes:
Exercise	1 2 3 4 5	
Food Quality	1 2 3 4 5	______________
Food Quantity	1 2 3 4 5	______________
DHA	1 2 3 4 5	______________
Mindfulness	1 2 3 4 5	______________
Attitude	1 2 3 4 5	______________
Sense of Curiosity	1 2 3 4 5	______________
Memory Stimulation	1 2 3 4 5	______________
TOTAL SCORE:	______________	______________
Blood pressure:	______________	______________
Pulse:	______________	______________

Week Twelve

This is your last week. Make it your best. Add intensity in one or more areas, or move up a track if you haven't already. As you finish this week and rate yourself for the last time, consider your next steps. If you've established solid habits—and changed your life—you can continue without the structure of my brain fitness program. If you'd like to continue following the program, simply use what you learned to create new challenges for yourself. There's no limit to your brain potential. Keep harnessing the power within—and enjoy the results.

Rate Yourself

Week 12

Date ____________

At the end of week twelve, please rate your efforts in the following areas.

		Notes:
Exercise	1 2 3 4 5	
Food Quality	1 2 3 4 5	____________
Food Quantity	1 2 3 4 5	____________
DHA	1 2 3 4 5	____________
Mindfulness	1 2 3 4 5	____________
Attitude	1 2 3 4 5	____________
Sense of Curiosity	1 2 3 4 5	____________
Memory Stimulation	1 2 3 4 5	____________
TOTAL SCORE:	____________	____________
Blood pressure:	____________	____________
Pulse:	____________	____________

Graduation!

Congratulations on completing your twelve-week brain fitness program. By now, you should be better able to remember names and faces and feel more mentally clear and creative. If I were to perform a brain MRI on you, I would likely see that your hippocampus is bigger and your cortex considerably denser. On EEG, your brain would likely be humming along in the focused, calm, and alert alpha range.

There's just one more step to seal the deal: your final Fotuhi Brain Fitness Score. Please rate yourself from 1 to 3 on the following brain fitness calculator:

Brain Fitness Calculator

		Score
LDL cholesterol	Give yourself a 1 if your LDL is high (more than 160 mg/dL), a 3 if your LDL is low (less than 100 mg/dL), or a 2 if it falls somewhere in between.	
HDL cholesterol	Give yourself a 1 if your HDL is low (less than 40), a 3 if your HDL is high (more than 60), or a 2 if it falls somewhere in between.	
Blood pressure	Give yourself a 1 if your average daily blood pressure reading is regularly 140/90 or worse, a 3 if it is regularly 120/80 or better, or a 2 if it falls somewhere in between.	
Vitamin B12	Give yourself a 1 if your B12 is less than 200, a 3 if it is more than 500 pg/mL, or a 2 if it falls somewhere in between.	
Vitamin D	Give yourself a 1 if your vitamin D is less than 40 nmo/L, a 3 if it's more than 60 nmo/L, or a 2 if it falls somewhere in between.	
Sleep quantity	Give yourself a 1 if you sleep fewer than six hours each night, have trouble falling asleep or staying asleep, or wake up extremely early. Give yourself a 3 if you fall asleep easily, get seven to eight hours of sleep each night, and wake up feeling rested. Give yourself a 2 if your sleep habits place you somewhere in between.	
Sleep quality	Give yourself a 1 if you snore loudly, feel very sleepy during the day, have no energy, and feel mentally foggy; a 3 if you never snore and do not have sleepiness during the day; or a 2 if you suspect you're somewhere in between.	
Weight	Give yourself a 1 if your waist-to-height ratio (WHtR) is 0.7 or more, a 3 if it's 0.5 or less, or a 2 if it falls somewhere in between.	

		Score
Activity level	Give yourself a 1 if you have a sedentary, "couch potato" lifestyle, a 3 if you get forty-five minutes of vigorous exercise five days a week, or a 2 if you're somewhere in between.	
Brain safety	Give yourself a 1 if you regularly engage in activities that can damage the brain (not wearing a helmet when cycling, skiing, horseback riding, or snowboarding; failing to wear a seat belt; boxing or practicing mixed martial arts), a 3 if you always protect your head or don't engage in brain-risky behaviors, or a 2 if you believe you fall somewhere in between.	
Social engagement	Give yourself a 1 if you lead a very isolated life and consciously avoid interacting with others, a 3 if you're a social butterfly and enjoy meeting new people, or a 2 if your social activity level falls somewhere in between.	
Alcohol use	Give yourself a 1 if you drink three or more glasses of alcohol daily or if you binge drink on weekends. Give yourself a 3 if you drink only socially or have on average one serving of alcohol a day (two for men). Give yourself a 2 if you fall somewhere in between.	
Food quantity	Give yourself a 1 if you overindulge, get second servings, and snack often; a 3 if you eat healthy portions with no more than one or two snacks a day; or a 2 if you fall somewhere in between.	
Food quality	Give yourself a 1 if your diet consists of fatty burgers, pizza, salty french fries, and sugary sodas; a 3 if your plate is usually like a rainbow, with a good mix of high-protein, low-carbohydrate, heart-healthy ingredients, and you eat fruits, vegetables, and nuts for snacks; or a 2 if your diet quality is somewhere in between.	
DHA	Give yourself a 1 if you never consume fish and do not take a DHA supplement, a 3 if you eat fish twice a week and take 1,000 milligrams of DHA daily, or a 2 if your DHA consumption falls somewhere in between.	
Mood	Give yourself a 1 if you are regularly sad, depressed, or irritable; a 3 if you are almost always cheerful, upbeat, and agreeable; or a 2 if your mood is somewhere in between.	
Mindfulness	Give yourself a 1 if you are stressed on a daily basis, a 3 if you rarely feel stressed and are often in the "alpha zone," or a 2 if your state of mind falls somewhere in between.	
Attitude	Give yourself a 1 if you are a pessimist, a 3 if you are an optimist, or a 2 if your attitude is somewhere in between.	

		Score
Memory stimulation	Give yourself a 1 if you never try to remember names or any other information; a 3 if you try to memorize the name of every person you meet and practice memorizing your grocery list, facts, or even poetry; or a 2 if your memory stimulation falls somewhere in between.	
Sense of curiosity	Give yourself a 1 if you rarely attempt to learn new skills, hobbies, or activities; a 3 if you try daily to learn new skills, hobbies, or activities; or a 2 if your sense of curiosity puts you somewhere in between.	
Total		

Now, look back to your beginning Fotuhi Brain Fitness Score from page 47 and record it below, along with your finishing score.

Beginning Score ____

Finishing Score ____

Your job in the weeks, months, and years ahead is to maintain or improve your score, living a brain-healthy life that will boost your cognitive performance today and into the future.

PART IV

Brain Shrinkers You *Should* Live Without

IF YOU THINK back to CogniCity, you'll recall that the metropolis will have plenty of wear and tear to contend with as the years pass. It has roads to repave, buildings to refurbish, parks to maintain, even sewage to dispose of. That's a lot to take care of. And that's the best-case scenario. The truth is that there is a good chance that CogniCity will also be the victim of one or more natural disasters over the years. Imagine what happens when a hurricane rolls through. Or a tropical storm. A snowstorm. A tornado. Or all four. If they were minor events, you might expect disruptions in traffic, difficulty conducting business, breakdowns in communication. But a major event—or the combination of several lesser events—could be catastrophic.

An equivalent scenario plays out in the brain. With age, our brains will slowly degrade as a natural result of "wear and tear." But add to that the effects of disasters, such as out-of-control obesity and diabetes, sky-high blood pressure, severe alcoholism, serious trauma, Alzheimer's disease, or a major stroke, and your brain will almost certainly suffer the consequences. Combine one or two of these and the effects multiply. The more you add, the greater the impact on the brain. Eventually, with enough disasters—large or small—you might find yourself the victim of serious shrinkage and brain frailty.

But even before then, your brain will feel the effects. You will think more slowly, forget more easily, experience a drain on your creative abilities, and be a poorer problem solver. Your mood may suffer and you may find yourself lacking a zest for life.

And, here's the kicker: you may not even realize it's happening. At least not for a while. If you're like the vast majority of my patients, you might worry vaguely about your memory or slowed thinking, but until symptoms become significant, you'll simply attribute any lapses to the unavoidable effects of aging or your busy life. In fact, most of my patients have no idea that their mental sluggishness has anything to do with health and lifestyle choices they make every day. Without even realizing it, they're doing their brains decades of disservice.

You already know about the brain's incredible potential for growth. Harnessing that power for change is a matter of striking a balance, every day, between brain growers and brain shrinkers. Every brain-healthy choice you make tilts the scale toward growth; every brain-unhealthy choice tips the balance in the opposite direction, with results that are felt both immediately and in the future.

A Web of Trouble

You're about to learn the major brain shrinkers we encounter in life. But it's important to know, as you think about how each affects your own brain, that brain shrinkers have a tendency to interact with each other to magnify the disaster.

Sleep disorders, for example, can contribute to other brain-draining conditions, including excess weight, stroke, and depression. As you'll read in the chapters to come, having one of these conditions may increase your risk of developing another brain-shrinking condition, which in turn may increase your risk of yet another brain-shrinking condition. This often sparks a powerful downward spiral of continued erosion and decline.

How much effect each will have is a matter of degrees: slightly high blood pressure may not shrink your brain, but years of uncontrolled, extreme high blood pressure will. The same applies to small versus large strokes, subtle versus severe sleep apnea, one minor hit to the head versus several large ones, and so on.

The silver lining in this web is that treating one condition may help to reduce your risk of the others as well. Treat them all and you'll be well on your way to turning that downward spiral on its head and riding the results all the way up to a bigger, better brain.

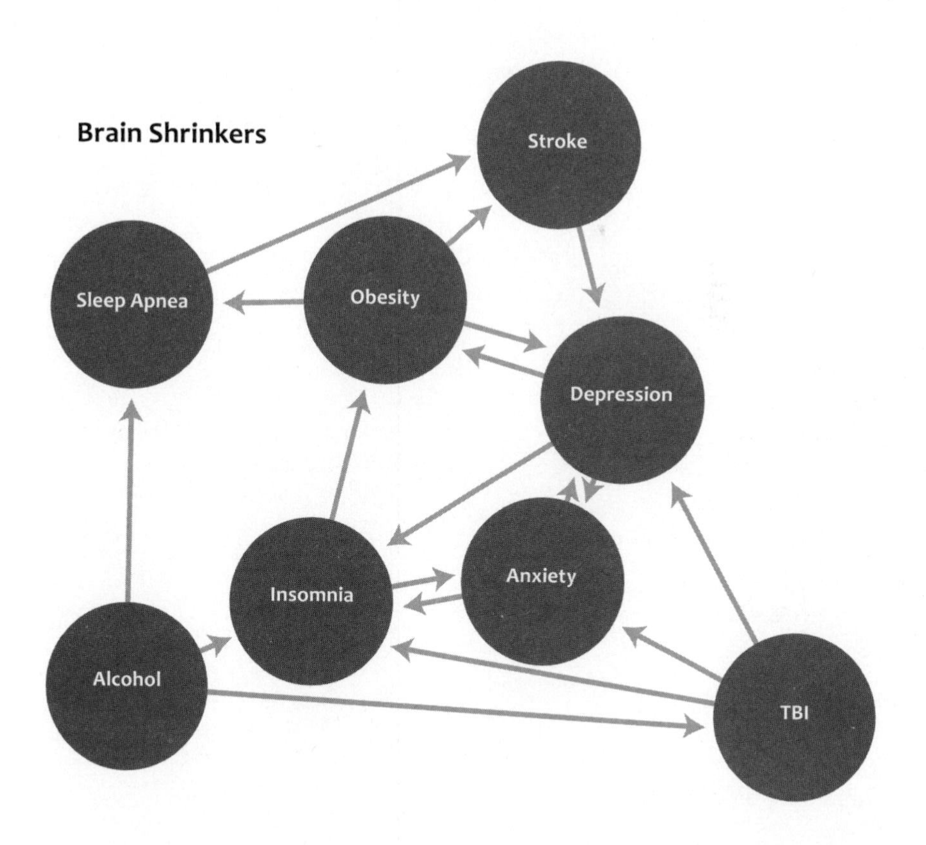

CHAPTER NINE

Shrinking the Brain One Night at a Time

IT HAPPENS THREE or four times a week. I'll ask a patient if she has health concerns and she'll rattle off a list: high blood pressure, a little extra padding around the waist, headaches, memory problems, fatigue. The one thing my patients often fail to mention is sleep. When I begin to prod them about their nighttime rest, I often hear the same lament. "Oh, I sleep terribly," they'll say. "I'm up till midnight and then I wake up at five A.M. for no reason. And I'm always tired!"

Often, it becomes apparent as we talk that they suffer from chronic insomnia, yet most have never sought treatment, nor do they realize their problem is almost certainly treatable. In fact, most of my patients who don't sleep well view their plight as a normal part of life. "Doesn't everyone have trouble sleeping?" they'll ask.

The answer is no. Or, rather, it should be no. In fact, insomnia and other sleep disorders are alarmingly common. The Centers for Disease Control calls insufficient sleep a public health epidemic, noting that fifty to seventy million Americans suffer from a sleep disorder.[1] But that doesn't mean we have to—or should—accept sleep disorders as a part of life. Indeed, doing so presents a real danger to our brains.

If you've ever suffered through a sleepless night, you know how it makes you feel the next day: tired, yes, but also foggy, irritable, and a little slow to react. Pull an all-nighter to prepare for a big presentation, for example, and before too long you'll feel the effects. Without rest, your body and brain lose out on a chance to rejuvenate—and the effect is a hard one to forget.

And while a less-than-stellar day may seem like the worst of it, prolonged periods of poor sleep are likely doing more damage than you imagine. In fact, serious—but fairly common—sleep disorders shrink the hippocampus and cortex, wither blood vessels, and erode the brain's highways. Even sleep habits that don't rise to the level of a disorder—like habitually staying up late and waking up early—may be robbing you of brain reserve.

I'm going to detail just how those brain shrinkers make their mark, but first you'll need to know just why and how we sleep.

A Good Night's Rest?

According to the National Institutes of Health, about forty million Americans suffer from chronic, long-term sleep disorders each year, with another twenty million saying they occasionally suffer from sleep disorders.

Understanding Sleep

When I visited Florence, Italy, in 2001 I was awed by its historic architecture. But there was something else I noticed, too: the streets and sidewalks were incredibly clean. It wasn't until I ventured out in the wee hours one night—ironically, thanks to jet-lag-induced insomnia—that I understood why. As most of the city slept, a fleet of trucks and city workers armed with brooms, dustbins, and even high-powered pressure washers, swept and scoured away the day's grime. When the sun rose, Florence awoke scrubbed and ready for a new day.

In a way, your brain—your CogniCity—is a lot like the city of Florence. By day, it's hard at work. By nightfall, it has accumulated the wear and tear—and metabolites, in the case of your neurons—to show for it. Your brain needs a good scrub. And while your body rests, your brain's maintenance crew gets to work, cleaning, rebuilding, and repairing.

Sleep is a fascinating process. Healthy sleep unfolds in a predictable pattern, cycling through periods that are categorized as non-rapid eye movement (NREM) and rapid eye movement (REM). NREM accounts for about 75 percent of your night's sleep and occurs in stages, during which you fall progressively deeper into slumber. During these stages your breathing and heart rate slow down, your body temperature drops, and your brain waves change from the short, spiky waves of an awake brain to longer, slower waves. During your deepest and most restful stages of sleep, your blood pressure drops and your muscles relax. It's in these stages of deep sleep that you find waking up most difficult.

NREM is followed by REM sleep, an active brain period during which dreaming occurs. Brain waves in REM sleep are short and choppy, similar to those in the awake brain. The body, on the other hand, is inactive (apart from the eyes, which dart under closed lids). You'll cycle through non-REM and REM sleep all night, with the REM stage becoming longer as the night progresses. As morning approaches, your level of cortisol rises, helping to insure alertness when you awaken.

Although optimal sleep varies by person, most people need an average of seven to eight and a half hours of uninterrupted sleep each night. So says Dr. Helene Emsellem, medical director of the Center for Sleep and Wake Disorders in Chevy Chase, Maryland, and a spokesperson for the National Sleep Foundation. One exception is teens, who need a little more shut-eye than adults—about nine and a quarter hours per night. There may also be rare exceptions for people with genetic differences; researchers have identified one gene whose carriers require as little as five to six hours of sleep a night.[2]

But for most people, as Emsellem says, seven hours of sleep is, "the lower legal limit." Unfortunately, it's not that unusual to find people whose work, school, or social schedules keep them regularly under that limit. And unlike animals, who "will just go to sleep" when their bodies need it, "humans are Energizer bunnies and will push themselves through on too little sleep," says Emsellem. They may go for long periods—years or even decades—shorting themselves on much-needed rest.

What they're missing out on is more than just repair time for the body and brain. In recent years, we've begun to discover just how important sleep is for memory, learning, and behavior. Sleep, it turns out, helps us consolidate the information we receive while awake.[3] In particular, the slow brain wave activity of certain stages of sleep seems to aid in hippocampus-dependent memory of facts and information, while REM sleep may help us process procedural memory.[4] Fragmented or poor-quality sleep, then, may interrupt or hinder the consolidation of memories for long-term storage.

Obstructive Sleep Apnea: Blocking Airflow with Brain-Shrinking Results

Although a host of sleep disorders exist, there are two I see most often in patients with memory complaints: obstructive sleep apnea (OSA) and insomnia.

We'll start with OSA, a disorder that is associated with dropping oxygen levels during sleep, followed by daytime fatigue and sleepiness, or a sensation of "foggy brain." One of the chief signs of OSA is heavy snoring, especially the kind that stops suddenly and is followed seconds later by a gasp for air. Those quiet moments are actually a dire sign: you're not breathing. Hence the gasp for air. That's not to say that all snoring is indicative of obstructive sleep apnea; some snoring is perfectly benign. OSA can be confirmed with a sleep study, which measures how many times per hour you experience a drop in your oxygen level. Fewer than five episodes per hour is acceptable, while fifteen episodes represent moderate sleep apnea, and more than fifteen represent severe OSA.

According to the American Sleep Apnea Association, some twenty-two million Americans suffer from OSA. But as many as 80 percent of OSA sufferers are undiagnosed.[5] And while it's long been considered a men's disease, we're now learning that OSA in women has probably been vastly underestimated. In one study of four hundred women, 84 percent of obese women had OSA.[6]

In fact, obesity is a major risk factor for OSA, in both men and women. In addition, having a large neck size—of more than seventeen inches in circumference for men or sixteen inches for women—increases your risk of OSA, as does having a small upper airway; a large tongue, tonsils, or uvula; a recessed chin, a small jaw, or a large overbite; and smoking or alcohol use. In addition, people over forty and African Americans, Pacific Islanders, and Hispanics are more likely to suffer from OSA.

The pauses in breathing caused by OSA can last from ten seconds to a full minute. Those lapses in breathing, of course, starve the brain of oxygen. Such

Stop Bang: Do You Have Sleep Apnea?

In 2012, researchers at the University of Toronto reported a high degree of accuracy for this clever OSA screening mnemonic:[7]

S for *snore:* Do you snore loudly? Loud enough to be heard through closed doors?

T for *tired:* Do you often feel tired during the day?

O for *observed:* Has your snoring been observed by someone else, such as a spouse?

P for *pressure:* Do you have high blood pressure?

B for *BMI:* Do you have a body mass index greater than 35?

A for *age:* Are you older than fifty?

N for *neck size:* Do you have a neck circumference larger than seventeen inches for men or sixteen inches for women?

G for *gender:* Are you male?

Answering yes to three or more of the above questions indicates a high risk for OSA.

oxygen restriction makes OSA, without question, one of the biggest brain shrinkers documented. In one study, patients with severe obstructive sleep apnea had brains with as much as 18 percent less grey matter than a control group.[8] I've seen time and time again how dramatic such a deficit can be. OSA patients feel forever tired, mentally foggy, and low on creativity. Plus they experience memory lapses that often send them my way in a desperate search for help. They may be irritable and even depressed as well.

Zapping Brain Growth

Why do my OSA patients have such trouble thinking? For starters, OSA starves the brain of oxygen, so much so that it raises the risk of strokes. Initially, sleep apnea can cause just small strokes, which may result in no major, noticeable problems, but they take their toll nonetheless. Such silent strokes, as you'll read in chapter 12, may slow a sufferer's thinking, cloud the memory, and even hinder the pace of walking or talking. In time, through its effect on the heart, on platelets in the blood, and on blood vessels, OSA can also contribute to major, disabling strokes.

In one study, presented at the American Stroke Association's International Stroke Conference in 2012, researchers found that of the people they studied who had silent strokes, more than half also had OSA.[9] And an incredible 91 percent of patients who'd had a large stroke also had OSA. In the same study, patients with OSA also showed damage to their white matter areas, the brain's network of highways.

This is due to the obvious effect that cutting off breathing has, but also to another factor that comes as a surprise to most of my patients. Sleep apnea causes blood platelets to become hypercoagulable, or more prone to clumping, which is one of the main reasons OSA contributes to an elevated risk of heart attack and stroke.

OSA is also associated with significantly low BDNF, as one study of twenty-eight men with OSA (compared to fourteen without) showed.[10] Being low on this brain fertilizer is a surefire way to miss out on brain growth and repair.

And OSA has a significant effect on brain wave activity. Often people with OSA have noticeable sleepiness during the day—a reflection of sluggish brain wave activity. A brain map EEG of an OSA patient, for example, would likely turn up too much slow theta activity, indicating the patient is far from the ideal alpha zone of the alert, calm, focused mind.

The end result is impaired function. In one telling study, Italian researchers followed a group of seventeen middle-age patients with severe OSA and fifteen volunteers without OSA, who acted as the control group.[11] All the study

The Gender Gap: Women Take a Bigger Hit?

OSA is more common in men than it is in women, but there's evidence that the damage done in women's brains may be greater than in their male counterparts'.

One study by researchers at the UCLA School of Nursing—the same group that first linked OSA to damage in the brain about a decade ago—found that women with OSA have greater damage to their brains' white matter, especially in areas of the brain important for mood regulation and decision making.[12] Women were more likely to be depressed as well (although I think it's important to note that OSA can raise the risk of depression in men, too).

subjects took a working memory test and underwent MRI scanning, including functional MRI scans that looked at brain activity, at the beginning and end of the study.

Before treatment, "even if they were not aware of their cognitive deficits, OSA patients were impaired on most cognitive tests," explains study author Vincenza Castronovo, a clinical psychologist and psychotherapist with the Sleep Disorders Center at Università Vita-Salute San Raffaele in Milan, Italy.

Functional MRI scans showed that when OSA patients performed a mental task, they "over-recruited" other brain regions in order to accomplish the task. In other words, they had to call on other parts of the brain in a sort of all-hands-on-deck surge of activity. "They were making more effort in order to perform," says Castronovo of the OSA patients.

Those with OSA also had smaller hippocampi as well as reductions of grey matter volume in the frontal lobes and white matter damage.

The effects of such damage may be amplified as we age, even as early as midlife. When researchers at the University of California, San Diego, compared fMRIs and cognitive tests of young and middle-age volunteers with OSA to those of young and middle-age volunteers without OSA, they found patients under the age of forty-five who had OSA were able to compensate by recruiting other areas of the brain to maintain performance—the all-hands-on-deck phenomenon at work again.[13] But people in the forty-five- to fifty-nine-year-old age group who had OSA displayed reduced performance for immediate word recall and slower reaction times during sustained attention tasks, indicating they weren't able to fully compensate. The researchers concluded that age combined with OSA offered a double insult—overwhelming the brain's capacity to respond to daily cognitive challenges.

A Subtle Start

When OSA sufferers come to see me, they often complain about troublesome cognitive problems. They know something isn't right; they just don't know what, or why.

But, as Dr. Castronovo demonstrated, structural damage in the brain may be quietly unfolding long before an OSA sufferer realizes it. In one study, researchers examined the MRIs of OSA patients and noted reduced grey matter volume in several brain areas, including the hippocampus, despite the fact that the study's sixteen participants performed fairly well on cognitive tests.[14] Eventually, with enough damage or with age, as noted above, their cognitive scores would likely decline.

Not only that, but their OSA would probably get worse. Why? The damage done by OSA can affect the area of the brain responsible for controlling breathing during sleep, increasing the severity of sleep apnea. As with many other brain shrinkers, it's a vicious cycle—the more you have sleep apnea the worse it may get.

Getting treatment sooner rather than later, then, is obviously the best bet. That advice is all the more compelling when you consider that damage from OSA to the brain's blood vessels may be measurable in mere weeks. In one animal model, for example, blood flow was restricted after just one month of OSA, a finding that is consistent with the reduced blood flow and oxygenation we see in patients with sleep apnea.[15]

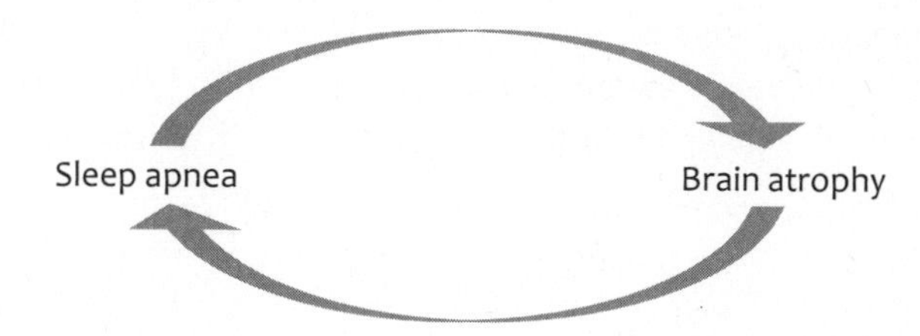

Insomnia: Shorting Your Brain on Sleep

In the winter of 2012 I began treating a patient, Gregory, who you might call a little type A. Gregory, a lawyer, habitually slept just four hours a night, dozing off at midnight and waking at four in the morning, without really even meaning to. Once awake, he'd fill the time with work. "It's not a problem," the fifty-year-old told me. "I'm used to it."

Except, it *was* a problem. Gregory had made an appointment to see me because he'd increasingly been experiencing memory lapses. They were the types of slips any busy person can relate to. Engrossed in work, he might forget to meet his wife for dinner. Or misplace a file. Or realize—while zipping down the highway—that he'd left his coffee cup on the car roof. It can happen to anyone. But lately it had been happening to Gregory quite a lot.

Like many people who have memory lapses, Gregory couldn't quite shake the fear that he was beginning to show the signs of Alzheimer's disease. After I completed my full evaluation, I was happy to tell him that his fear was unfounded. But that didn't mean he was off the hook. Gregory's lifestyle, I explained, was unquestionably shrinking his brain.

Like many insomnia sufferers, he had never talked to a doctor about his early-morning awakening, considering it just a quirk that he had to live with. Had he known the damage he was doing to his brain, though, he might not have been quite so cavalier.

First, a little about the condition, which is one of interrupted or inadequate sleep. Insomnia is defined by a difficulty to initiate or maintain sleep for at least one week. If you suffer from insomnia, you may have trouble falling asleep, may fall asleep but awaken at night and have trouble getting back to sleep, may awaken very early or wake up not feeling refreshed. Whatever the case, you're not getting long stretches of uninterrupted sleep, which means you're not benefiting from a good night's slumber.

Insomnia is often related to or associated with another problem, either physical or psychological. Stress, depression, medications, caffeine, alcohol, shift work, chronic pain, or health problems (such as the hormonal fluctuations of menopause or restless leg syndrome, which could be due in part to iron-deficiency anemia) all can cause insufficient restful sleep. Often insomnia results in irritability and difficulty with attention, memory, and concentration.

What's Going On in the Brain?

In the short term, insufficient sleep triggers your sympathetic nervous system, which, as you'll read in chapter 10, has an immediate effect on your brain function. Cortisol, released during this response, heightens awareness—it's part of the fight-or-flight response you need to respond to emergencies. But the brain on excess cortisol isn't as accurate in solving complicated problems, making it more likely that you'll make mistakes in the heat of the moment. High cortisol can also affect your brain wave activity—insomnia patients have higher levels of excessively fast beta activity, a sign of hyperarousal in the brain.[16]

In the long term, poor or inadequate sleep has a direct effect on the structure of the brain. Animal studies have shown that sleep deprivation is tied to a smaller hippocampus, and one study of thirty-six men showed the same effect in humans.[17] There's evidence, too, that the increase in cortisol that accompanies insufficient sleep inhibits neurogenesis in the hippocampus.[18]

Insufficient sleep also affects brain structure and health indirectly. Regularly getting fewer than six hours of sleep a night is associated with chronic pain, hypertension, and inflammation, all of which may affect the brain. And researchers at the University of Alabama at Birmingham concluded that fewer than six hours of sleep a night significantly increases the risk of stroke for people who aren't overweight or obese and don't have OSA.[19]

Shorting yourself on sleep also may raise your risk of obesity, a brain shrinker you'll read about in chapter 11. And obesity, as you know, increases the risk of OSA. In fact, between 30 and 40 percent of my patients have both sleep apnea and insomnia, a double whammy for the brain. Insomnia is also one of the chief diagnostic features of depression, which takes its own toll on the brain and on patients' lives.

Brain function, not surprisingly, takes a hit from prolonged sleep problems. In one study of more than fifteen thousand nurses over the age of seventy, those who had fewer than five hours of sleep per night over a six-year period did worse on cognitive assessments than those who slept an average of seven hours a night.[20]

The Sleep–Food Link

Poor sleep may be affecting you in another way that shrinks the brain: lowering your resistance to foods you might otherwise chose to avoid. In one study conducted by researchers at the University of California, Berkeley, healthy adults deprived of sleep for a night were more likely to experience heightened desire when shown a food item than they were after having a full night's sleep. Functional MRIs of the twenty-three subjects who were sleep deprived also showed decreased activation in the frontal lobes—the brain's braking system—just the place they'd need to recruit to make healthy food choices.

In another study, researchers found that a single night without sleep resulted in increased feelings of hunger in twelve healthy young men.[21] In addition, when viewing images of food, the men showed greater activation in the right anterior cingulate cortex, an area of the brain tied to desire and craving. Activation here might explain their urge to eat.

Getting Better: Reversing the Damage of Sleep Disorders

If you're worried about your snoring, or the late nights and early mornings that are your current way of life, that's probably a good thing. OSA and insomnia are major brain shrinkers. Stopping further damage as soon as possible is critical to ensuring your brain functions at its peak, now and in the future. This is so critical, in fact, that I recommend sleep apnea screening for anyone over the age of fifty who snores. I think it is as important—if not more so!—than getting a colonoscopy or mammogram at this age. Doctors should routinely include questions about sleep problems in the evaluations of their patients.

If you have OSA or insomnia, you should make a commitment to starting treatment immediately, and sticking with it.

Patients with OSA should avoid drinking alcohol at night and taking sedating medications; these can worsen OSA. They should also be vigilant about reducing their other stroke risk factors, since having OSA ups their risk of stroke. The good news is that OSA can be very effectively treated with a continuous positive airway pressure (CPAP) device, which blows air into your nose while you sleep, keeping your airway open. Often, after a few days using CPAP, patients report feeling better than they have in years. Other treatments include mouth appliances that keep the airway open, surgery (for anatomical causes of OSA), and conservative treatments like sleeping on your side. There are also medications that can help to reduce daytime sleepiness in patients. I usually give a short trial of modafinil to my patients for a few days to see if they improve. If they do, they can take it regularly. Many times OSA also resolves completely with weight loss.

And there's more good news: as terrible as sleep disorders are for the brain, much of the damage they cause seems to be reversible. The results can be stunning. Remember the brain-shrinking effects of OSA reported by the Italian research team? Their research was actually aimed not just at measuring the impact of OSA, but also at testing the theory that treatment of OSA would result in significant improvements in brain structure and performance on cognitive tests.

After initial testing of the study subjects, Castronovo and her team treated the patients with CPAP for three months and then tested and imaged them again. With three months of treatment, fMRIs of OSA patients—which, remember, had shown them to be working overtime to complete cognitive tasks—now looked very similar to those of non-OSA study subjects. Treated OSA patients also did better on cognitive tasks and showed an increase in the size of the hippocampus and frontal lobes compared to before treatments. White matter in OSA patients had also largely normalized. If you're not already amazed by the incredible plasticity of the human brain, this alone will surely convince you.

More study is needed, but it seems there is hope that just as the lungs recover over a period of years after a person stops pumping them with cigarette smoke, so too the brain recovers and rejuvenates once the brain-draining effects of sleep apnea are eliminated.

In children, at least, this seems to be true. Children who had sleep apnea caused by large adenoids or tonsils, for example, had poor sleep quality and did poorly on cognitive tests compared to a non-sleep-apnea control group. But their BDNF levels, sleep quality, and cognitive abilities all had risen six months after having surgery to eliminate the cause of sleep apnea.[22] After twelve months BDNF levels, sleep quality, and cognitive abilities were no different in the children who once had sleep apnea than they were in the children who'd never had it.

The brain-draining effects of insomnia likely aren't indelibly etched either. In one animal study, for example, two weeks of treatment with a sleep aid resulted in a 46 percent greater survival of newborn hippocampal neurons.[23]

That's not necessarily a prescription for success, however. I usually use sleep aids as a last resort or a temporary measure for my insomnia patients since medications can be habit forming and long-term use can be associated with increased risk of cognitive impairment with aging. Therefore, I encourage my patients to gear their efforts toward eliminating the root causes of insomnia and incorporating brain-boosting activities into their lives. If work stress is keeping you up at night, for example, using the ABC method outlined in chapter 6 may help you reduce your anxiety so you can slumber soundly. Meditation and exercise (but not right before bed) should also improve your sleep.

Brain Meter

Association with Brain Shrinkage

Untreated sleep problems, especially obstructive sleep apnea, are associated with significant risk for strokes, along with brain atrophy. The worse your sleep disorder, the more your brain will shrink.

Cashing In Your Sleep IOU

If you make a habit of giving short shrift to your sleep, it's tempting to think you can "catch up" with a marathon snooze session once whatever is keeping you up has passed. But can you really?

Scientists don't yet have a solid answer, says Dr. Emsellem, but it looks likely that "catching up" on sleep may not be as simple as you think. Reversing the effects of a serious sleep debt "may take as much as four to six weeks of sustained sleep to really balance out and catch back up," she warns.

CHAPTER TEN

How Stress or Depression May Be Shrinking Your Brain

YOU'VE BEEN working on the proposal for months, staying late at work, and eschewing weekend downtime in favor of long hours at your computer. If you nail this presentation, your company stands a good chance of winning a key contract and avoiding layoffs, including—quite likely—your own. You know you've done a bang-up job, but as presentation time approaches, the pressure is on. And, honestly, you're not exactly feeling in peak mental shape.

In fact, you're beginning to feel like you're losing your marbles. Case in point: you just spent twenty minutes looking for the printout you'd had in your hand but then absentmindedly laid down (on top of the filing cabinet, as it turns out). And you'd left the office for a quick bout of errand running but then forgot to make two essential stops. At home, you've been forgetful and fuzzy. Are you just imagining things or has your formerly reliable brain gone haywire—just when you need it most?

Chances are you really are experiencing slower thinking and increased forgetfulness. And there's a perfectly good explanation: stress changes the brain—subtly with small doses; more dramatically with prolonged or extreme stress; perhaps most dramatically of all with chronic, major depression. Prolonged excess stress, we now know, causes the parts of the brain responsible for memory, attention, and decision making to shrink.

Too much stress can tip some people into depression, which also shrinks the brain. If you're feeling a little tense just thinking about this, take a deep breath. The damage caused by stress and depression is reversible. We'll get to that, but first you'll need to understand just how stress damages your brain.

(A Little) Stress Is Good

As you probably know, you're equipped with a rather handy stress-response system, which is often referred to as your fight-or-flight response. In fight-or-flight mode, your sympathetic nervous system releases hormones, including cortisol, which works with adrenaline to increase blood flow to your heart and other muscles throughout your body, and to heighten your awareness. In other words, it prepares you to either defend yourself (fight) or run (flight).

This emergency response system is designed to work for as long as it takes you to reach safety—from mere seconds to hours. When the emergency has passed, your parasympathetic nervous system releases hormones that counteract those released in fight-or-flight mode and your body and brain return to a normal state. Your heart rate slows; your muscles relax.

Being exposed to some stress can be helpful—and not just when you're trapped in a burning building or staring down a predator. In fact, mild, "acute," or short-term stress actually heightens your attention and focus. Consider the stress you feel when you sit down to take an exam or be interviewed for a job. Most likely it's just powerful enough to help you focus on the task at hand—and do your best. This is the fight-or-flight mechanism working as intended, giving you just enough energy and focus to perform at your peak.

To Stress or Not to Stress?

You probably know someone who becomes irate when stuck in traffic. Or who breaks into a cold sweat at the thought of speaking to a large group of people. We all respond to perceived stressors in different ways. Just how we handle them depends on a variety of factors, including some of our earliest life experiences. We know this, in part, thanks to the research of McGill University professor Michael Meaney, an expert on stress and the brain, whom I worked with when I was an undergraduate student. It was working with Dr. Meaney, in fact, that sparked my long-lasting interest in how stress affects the brain.

One of Dr. Meaney's best-known studies looked at what happened to infant animals who were cared for—or neglected—by their mothers. As it turned out, baby animals groomed by their mothers experienced changes in their brains that made them better able to handle stress later in life. Animals who were denied maternal care early in life, on the other hand, were less able to handle stress once they reached adulthood. Early life experiences affected their stress response later in life.

But there's also learned behavior around our responses to stress. Part of it we pick up from our parents, whose reactions to events in their own lives give us cues—for better or for worse—on how to respond when it's our turn.

Much of the time, the stress we feel is our own doing. How we respond to it has to do, as you'll recall from chapter 6, with our belief system. It's all in how we view the world. A simple example is this: My co-author, Christina, might care deeply about the outcome of a game being played on a particular fall Sunday in the Baltimore Ravens' stadium. She's a Ravens fan, so a win matters to her. But at the very same moment that she is standing in front of the TV, biting her nails over a crucial first down, another person on the same block might feel no stress whatsoever. To him, a completed pass by Joe Flacco matters not at all. Therefore, he is utterly relaxed at game time.

This is a rather benign example, but imagine if the stressor were a poor performance review at work, or a flat tire, or a nagging relative. Whether we consider these mild annoyances or major stressors depends largely on our priorities, our life circumstances, and our belief system.

Excess Stress Is Bad

While some stress is helpful, *excess* stress can have negative effects on performance. To demonstrate this, researchers at the National Institute of Mental Health looked at the immediate effect of a stressor on healthy subjects.[1] For the study, the research team recruited twenty-four volunteers and divided them into two groups. Study subjects in one group were exposed to stress by having one hand dunked into ice water; in the other group, participants had one hand dunked into warm water. Twenty minutes later, both groups were tested on working-memory performance and their heart rates and salivary cortisol levels were measured. It turns out that those exposed to short-term stress—the cold water—had better reaction times on the test but also had more false alarms and made more mistakes than the control group. In fight-or-flight mode, in other words, they were fast but not so accurate.

This was a short-term effect, of course. The real trouble with stress shows itself when we're exposed to too much, for too long. In fact, a host of studies have shown a negative effect on the brain from prolonged exposure to excessive stress.

In one such study, researchers exposed rats to a cat over a period of five weeks and then tested the rats' memories using a puzzle-solving maze that required them to remember the location of a platform. Non-stressed rats did well on the test; their stressed-out peers consistently failed.[2]

In another study, researchers at Washington University School of Medicine in St. Louis gave volunteers oral doses of cortisol that approximated the levels they'd experience under stress. The subjects were given a memory test before being given cortisol and then after one day of cortisol, after four days of cortisol, and six days after they'd stopped taking it. They were compared to a control group not given cortisol.

What happened? Although cortisol seemed to have no short-term effect on nonverbal memory, sustained or selective attention, or executive function, those given cortisol did worse on tests of verbal declarative memory. The more cortisol they'd had, the worse they did.

Longer term, excessive stress creates even bigger deficits. Why would stress have such a negative effect on brain function? For starters, "chronic stress causes at least some nerve cells in key brain regions like the hippocampus to shrink and lose their connections with other nerve cells," explains Dr. Bruce McEwen, a professor at The Rockefeller University, who for decades has studied the effects of stress and sex hormones on the brain (and who was a postdoctoral mentor for my undergraduate professor, Dr. Meaney). The process is called synaptic degeneration, and the result is a breakdown in communication between brain cells.

Excess stress has also been shown to inhibit neurogenesis in the hippocampus. "As a result, the hippocampus becomes smaller and less effective in doing its job," says Dr. McEwen, whose animal studies have helped shed light on the effects of stress on the hippocampus. Human studies have shown similar effects on the hippocampus.[3]

Over time, the effects accumulate. In one study, women who reported a history of perceived high stress over twenty years had smaller hippocampi than those who reported less stress.[4]

One reason for that shrinkage may be that stress lowers BDNF levels. In one animal study, rats who were intentionally stressed for six hours had reduced BDNF levels in their hippocampi. Another study, this one in humans, found reduced BDNF levels in patients who'd had childhood trauma and recent stress.[5] Reduced BDNF and increased cortisol levels were also tied to a smaller hippocampus. It makes perfect sense: as BDNF is repeatedly lowered, new neurons aren't able to survive. That, combined with synaptic degeneration, leads to hippocampal atrophy.

When it comes to chronic stress, however, the hippocampus is not all we have to worry about. In recent years evidence has emerged showing that stress does damage to other parts of the brain, most notably the prefrontal cortex—a key area for working memory, decision making, executive function, and impulse control. The prefrontal cortex also works in ways that affect behavior, for example, by helping to keep the amygdala—a key center for negative emotions—under control.

Animal studies have shown that chronic excess stress causes neurons to shrink and lose connections and functional efficacy in the prefrontal cortex, "just like in the hippocampus," says Dr. McEwen, whose many studies included one on medical students preparing for their board exams. In that study, Dr. McEwen found that students' perceived level of stress was tied to poor performance on a task that relied on the prefrontal cortex.[6] The more stressed they felt, the worse they did. Connectivity in the prefrontal cortex, as seen on fMRI, was also reduced with high stress.

Meanwhile, though neurons in the prefrontal cortex are shrinking and showing disconnection, neurons in two other parts of the brain may actually grow with stress. One of these growth areas is the orbital frontal cortex, which responds to stimuli that predict reward or punishment. The other growth areas are in segments of the amygdala that play a role in fear, anxiety, and aggression. In animal studies, growth in these areas makes the animal "more anxious and more fearful and, in the case of the orbital frontal cortex, more sensitive to stimuli that either predict reward or punishment," explains Dr. McEwen. Growth in these areas of the brain serves a purpose: if you're a deer heading out across the tall grass of an open field—where predators can lurk—heightened awareness and a little anxiety might save your life. But when the danger passes, "if these changes do not reverse, then you have an anxiety disorder," says Dr. McEwen.

Think You're Not Tense?

Some stressors—an overdraft notice from your bank, a fight with your spouse, a mistake at work—are hard to miss because they elicit an immediate response: you feel your heart rate increase, your emotions rise.

But you may be subjected to stressors without even noticing them. One possible culprit is noise. And not just loud, jarring noises. Our brains and bodies actually respond to noise that we don't even realize we hear. In an evolutionary way, that makes perfect sense. If your fight-or-flight response didn't kick in when your brain picked up the almost-silent rustle of leaves in a nearby bush, you might find yourself in the jaws of a predator. In a true emergency, a keen sense of hearing—and your corresponding fight-or-flight response—still comes in handy. But in cases where there's no real emergency, noise can simply create stress. And chronic exposure to noise can create chronic stress.

That means the workspace floor plan that's supposed to encourage collaboration could actually be sending your stress level soaring. In 2000, environmental psychologist Gary Evans and a colleague at Cornell University reported that working in a moderately noisy open-format office results in higher levels of stress.[7] That study compared stress levels of twenty workers randomly assigned to a quiet office and twenty workers subjected to three hours of noise intended to approximate the low-intensity sounds of an open-format office. Once the workers were in place, the research team performed tests to measure stress hormone levels and cognitive skills. They found that those in the noisy office had elevated stress hormones and made fewer attempts at unsolvable puzzles. That was true even though the workers didn't report feeling higher levels of stress than their counterparts in quiet offices.

Even low-level noise like nearby traffic can cause stress, as a Cornell team found in a study of 115 Austrian fourth-graders.[8] Those who lived in noisier residential areas had slightly higher resting systolic blood pressure, greater heart reactivity when presented with a stressor, and higher overnight cortisol levels, compared to children who lived in quiet residential areas.

Chronic exposure to louder noise, like that from an airport, can have a significant impact, too, as a variety of cross-sectional studies have shown in the past. In the late 1990s the closing of one airport and opening of another in Germany allowed researchers to see for the first time what happened when children switched from being in an area that is quiet to being in an area that is loud, and vice versa.

The study, conducted by Evans and counterparts in Germany and Sweden, followed 326 third- and fourth-grade children living in Munich, half near an existing airport that was slated to close and half near the site of a new

Stress Nation?

When the American Psychological Association (APA) surveyed Americans about their stress levels in 2011, the picture wasn't exactly bright. Some 22 percent of those surveyed said they were under extreme stress, ranking their stress level an 8, 9, or 10 on a scale of 1 to 10. The survey probably undercounted the highly stressed, too, since people under serious duress may be too pressed for time to answer a survey. Based on the thousands of patients I've seen, I'd estimate the true amount is closer to 50 percent.

Interestingly the survey also noted that 36 percent of respondents thought their stress levels had no bearing on their mental health. Only 29 percent felt they were doing an excellent or very good job at managing or reducing their stress levels.

airport, soon to open. The children were tested six months before the new airport opened and again one and two years after it opened.[9] The researchers found that, after the new airport opened, children in the high noise areas near it showed impairment in long-term memory, reading, and speech perception. Those in the now quiet area near the old airport, meanwhile, saw improved performance in long-term memory and reading as well as in short-term memory.

The fact that scores improved once noise was removed is a promising sign that the ill effects of noise pollution are likely reversible.

Depression: A Brain Changer

Depression is a mood disorder characterized by prolonged feelings of sadness and other key symptoms, including disinterest or lack of pleasure in daily activities, insomnia or excessive sleepiness, excessive weight loss or weight gain, lack of energy, an inability to concentrate, feelings of worthlessness or excessive guilt, and/or recurrent thoughts of death or suicide. Depression can have many causes, including a genetic component that may make some people more prone to developing it than others.

Obviously depression isn't merely a matter of being excessively stressed. But prolonged stress—especially the type caused by factors beyond our control—can play a part in the development of depression. And feeling perpetually stressed, or having persistent anxiety, can also be a symptom of depression. In my own medical practice I often see patients who are depressed but don't realize it; they instead attribute their symptoms to stress. Often they're surprised when I suggest that they may meet the criteria for clinical depression.

But they're certainly not alone. In fact, the Centers for Disease Control (CDC) reports that 9 percent of the American adults it polled reported suffering from major depression in the two weeks before being polled. Many others may suffer from lesser depressive symptoms.

Their symptoms aren't terribly surprising when you consider what's going on in the brain. For starters, depressed people have lower levels of norepinephrine and serotonin, two key neurotransmitters that help pass messages in the brain, especially in areas important for attention, mood, reward, and alertness. Without the full support of these messengers, depressed brains don't operate as well.

High cortisol levels have also long been noted in people with depression. And since excess cortisol is tied to a smaller hippocampus, it's no surprise that depression is associated with shrinkage in the hippocampus as multiple studies have shown. Although there are some exceptions—one large study

Who Is Depressed?

According to the CDC, those most likely to report suffering from depression are:

- forty-five to sixty-four years of age;
- women;
- blacks, Hispanics, persons of other non-Caucasian races or of multiple races;
- people with less than a high school education;
- people who have been previously married;
- unable to work or currently unemployed; and/or
- without health insurance coverage.

failed to find this link—there is ample evidence that the worse the depression, the smaller the hippocampus. One meta-study that looked at the results of twelve other studies found that people with major depression may have hippocampi that are, on average, 8 to 10 percent smaller than people without depression.[10]

That size difference doesn't tell us which came first, of course. We don't know for sure whether depression shrinks the hippocampus or whether people with smaller hippocampi are more prone to depression. One MRI study of children with a family history of depression found that the children had smaller hippocampi even before they showed any signs of depression.[11] Other studies, however, show conflicting evidence.

It's quite possible, given the contradictory evidence, that the answer to the question might be both: depression shrinks the hippocampus and a smaller hippocampus makes people more prone to depression. We do know, however, that depression is linked with lower BDNF, which may explain that smaller hippocampal size.[12]

And while no study I know of has found that depression *causes* reduced blood flow to the brain, the symptoms of depression—like feeling blue or lacking an interest in life—may involve sluggish blood flow to the brain. This is exactly what one study showed.[13] Researchers at the Royal Free Hospital's School of Medicine in London used PET scans to measure blood flow in thirty-three patients with known depression. They found that melancholic patients had reduced cerebral blood flow in the prefrontal and limbic areas. Other studies have found significant decreases in blood flow to the prefrontal cortex.[14]

Depression also, not surprisingly, puts a serious damper on healthy brain activity. The EEG brain map of a melancholic depressed patient, for example, might show too much slow theta activity on one side of the brain. If you imagine the brain as an orchestra, you can think of this orchestra being comprised of a collection of tubas producing mostly low and slow sounds. A person with an anxious type of depression, on the other hand, might produce a map showing too much high beta, particularly on the right side of the brain. The anxious person's orchestra would be made up primarily of violins playing screechy high notes.

Both maps represent brain activity that's far from the calm, focused alpha zone we know to be vital for a strong brain. It's no surprise, then, that depressed patients report being forgetful, unable to pay attention, and lacking in clarity and creativity.

Depression is also interlinked with other brain shrinkers, such as obesity and insomnia. The upside of that connection is that treating one condition may help improve the others.

Reversing the Effects

Stress is so pervasive, and the harmful effects on the brain so profound, that I believe treating and preventing it should be considered a national priority. And I'm not alone in thinking that. My friend and colleague, Cleveland Clinic preventive medicine professor Dr. Michael Roizen, views stress management as more critical even than stopping smoking. Dr. Roizen is a co-author of the

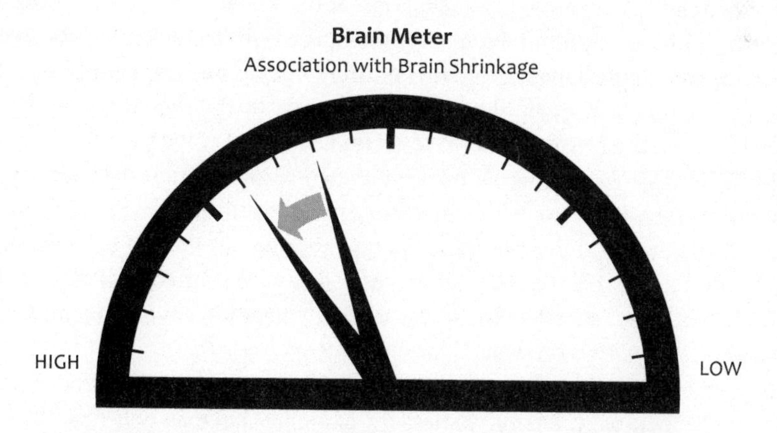

Chronic and excessive stress, especially when combined with major depression, insomnia, or high blood pressure, is associated with brain atrophy. The worse the symptoms and the longer they persist, the more your brain will shrink.

Extreme Stress: PTSD

At the extreme end of the stress spectrum is post-traumatic stress disorder (PTSD), an anxiety disorder that occurs when someone is exposed to a highly stressful event. People can suffer PTSD even if they aren't physically harmed by the event or are only a witness to the event.

Sufferers of PTSD experience an array of psychological problems, from persistent anxiety and vivid memories of the traumatic event, to avoidance of anything that reminds them of the event, to sleep problems and an inability to stop thinking about the event. Sufferers might also become emotionally detached, numb, or even self-destructive.

Scientists are still trying to understand exactly what happens in the brain to cause PTSD. Why do some develop it and others do not? We still don't know. But we do know that people with PTSD have smaller hippocampi.

One possible explanation for this is that people who develop PTSD actually had smaller hippocampi to begin with. In one study of identical twins, for example, researchers found smaller hippocampi in people who'd been in combat and suffered PTSD *and* in their identical twins, who hadn't been in combat and didn't have PTSD.[15] That finding seems to suggest that having a smaller hippocampus may predispose a person to developing PTSD if that person is exposed to a traumatic event. However, more research needs to be done to better understand PTSD.

YOU series with Dr. Oz, and knowing all the different factors that affect our bodies and minds, he puts stress reduction at the top of his list of priorities for improving public health and reducing the burden of diseases in our country.

Fortunately, stress management is entirely doable and depression is treatable in most people. Even better, when it comes to the damage wrought by chronic stress, the effects do seem to be reversible.

In the case of Dr. McEwen's medical students, for example, one month after they'd taken their exams, connectivity problems in the brains of formerly stressed students had disappeared. "So, in these young human subjects these kinds of effects of everyday experiences of feeling out of control of your life actually can have reversible impairments that you can recover from by taking a vacation," says Dr. McEwen.

Even the brain shrinkage associated with depression can be largely reversed with medical therapy. Many studies, for example, have shown that the hippocampus begins to grow again after treatment of depression. Antidepressant medications, in particular, have been tied to such changes. One study of twenty-four patients with major depressive disorders, for example, found that depressed people treated with antidepressants had increased angiogenesis as well as increased neurogenesis in their hippocampi, compared to untreated depressed patients.[16]

The Aging Brain

The research on the reversibility of brain changes caused by stress is encouraging. But before you heave a sigh of relief, you should know that older brains might not recover from brain changes quite as easily as younger brains. Older animals, for example, show the same response to stress as younger animals, but their brains don't "bounce back" after a stress-free recovery period in the way that younger brains do, says Dr. McEwen. This may be a matter of brain reserve—after a lifetime of wear and tear, an older person has less of a buffer zone to work with. An insult that might barely make a dent on a younger person's brain might tip an older counterpart closer to cognitive decline.

Certainly, the damage done by stress and depression appears to increase the risk of dementia later in life. Both shrink the hippocampus. And one study found depressed people over the age of 65 experienced 24 percent greater annual cognitive decline than their non-depressed peers.[17]

There is evidence, too, that stress may contribute to the risk of developing vascular dementia, a type of dementia caused primarily by small and large strokes. In a 2011 study, researchers evaluated 257 demented people and more than 9,000 non-demented people who'd taken part in the Study of Dementia in Swedish Twins.[18] They found that having a stressful job with low control, along with low social support, increased the risk of developing dementia, particularly vascular dementia. This is most likely related to the link between high stress and high cortisol, leading to high blood pressure and stroke. Making changes in the workplace so that employees feel they play a more meaningful role might help to lower that risk, the study authors suggested.

Laughter as Medicine?

As a medical intern assigned for two months to the coronary care unit at the Johns Hopkins Hospital in the 1990s, I quickly became accustomed to patients who weren't in tip-top shape. After all, most of the patients there had suffered heart attacks and were in the early stages of recovery. But there was something that stood out in my mind about the heart disease patients I encountered: they were very often, although not always, overly tense individuals. These were hard-charging executives, lawyers, and others who worked long days, which left them little time for fun. And they were, well, grumpy. Grumpier than any other group of patients I came across, quite honestly, even in the oncology, surgical, or gastrointestinal units. Were they unhappy because they were in the hospital, or were they in the hospital because they were unhappy?

I'll never know the answer, but those patients always pop into my mind when I talk to my own patients about the importance of living a balanced life and the value of laughter and happiness—even with a busy schedule.

Science backs me up on this. In 2001, researchers at the University of Maryland Medical Center (UMMC) reported their study on the benefits of laughter in heart health.[19] For the study, the team used questionnaires to gauge how prone to laughter three hundred volunteers were. Tallying the results, the research team found that those who'd suffered a heart attack or undergone coronary artery bypass surgery responded with less humor to everyday problems compared with those who had normal heart health.

Could being good-humored actually improve heart health? The study didn't preclude the possibility that heart problems might make people less likely to laugh (rather than less laughter contributing to heart disease), so the research team, led by the director of UMMC's Center for Preventive Cardiology, Dr. Michael Miller, tried another angle.

This time they studied twenty volunteers in good cardiovascular health, measuring the flow of blood through the arteries of their upper arms at baseline and then again just after watching a funny movie clip and just after watching a violent movie clip.[20] They found that laughter appeared to increase blood flow. The violent movie clip, on the other hand, caused volunteers' blood vessels to constrict, reducing blood flow.

Just why that happened isn't yet clear. But other studies have given more cause to engage in a hearty chuckle. Some have found that laughter reduces the level of cortisol and boosts our immune systems.[21]

That's not to say you need to laugh constantly; what's important is maintaining a happy mood and a good sense of humor.

Getting Better

There are countless strategies for reducing stress in our lives, as a quick glance at any of the many health magazines on newsstands bears out. I often find that the biggest obstacle for my patients isn't learning what to do; it's acknowledging they have a problem to begin with. Many simply don't believe their stress level is harming their brains.

They have me as their doctor, though, and I usually convince them rather quickly of the importance of stress reduction as well as the treatment of depression. If you suspect stress or depression may be an issue for you and you're not already talking to your doctor about it, you should do so immediately. This is especially important in the case of depression, since depression symptoms can actually be caused by health problems, such as low levels of B12, thyroid problems, anemia, sleep apnea, and silent strokes, among other things. Certain medications, especially pain medications, can also contribute to a depressed mood.

At my Brain Center, although I do prescribe antidepressants when needed, I also focus on eliminating symptoms through other measures. One is sleep improvement. Often depression goes hand in hand with sleep disorders, so I work both angles to try to reduce the symptoms of each. Another is exercise, which has been shown to be incredibly effective—some suggest as effective as antidepressants—in ameliorating the symptoms of depression. Often my patients have to start slowly with exercise. Depression affects their energy level, so even something as seemingly simple as a five-minute walk may be overwhelming for them. I have them start small and then work their way up to longer, more rigorous exercise as their depression symptoms improve. You already know exercise's brain-boosting potential, so you won't be surprised to hear that, often, the more they exercise the better they feel.

To help reduce stress and anxiety, my patients also use a combined meditation–cognitive behavioral therapy approach that I described in chapter 6. Using the ABC method, they try to change the way they view and respond to stressors in their lives—and improve their mood and attitude as a result.

And while no supplement has been shown to treat depression, healthy eating has clear brain benefits, so I prescribe for all my stressed and depressed patients the dietary changes you read about in chapter 5. Exercise and dietary changes have many other benefits, including helping to reduce the risk of obesity.

Patients enrolled in my brain fitness program also undergo neurofeedback and cognitive training to help promote healthy brain activity and improve their attention, focus, and problem-solving abilities. Last but certainly not least, I strongly recommend that all my patients find natural ways to boost their mood every day. Whether that's finding a favorite go-to website that's guaranteed to give them a chuckle, having a smile-inducing photo album handy, or finding some other form of instant cheer, laughter is good for heart health, which of course benefits the brain.

CHAPTER ELEVEN

Bigger Belly, Smaller Brain

YOU PROBABLY won't be surprised to hear that I'm fairly careful about eating, exercising, and maintaining a healthy weight. I have, after all, spent the past twenty years studying the effects of such things on the brain, and what I've learned has shaped my behavior. But I wasn't always health conscious.

In fact, as a medical intern I developed a set of lifestyle habits that left me, for a short while, with a bit of blubber around the belly. Most weeks I clocked more than a hundred hours of work in a high-stress environment, leaving me little time for exercise, a full night's sleep, or a healthy meal. Often I'd eat whatever was handiest—a slice of pizza, a cafeteria cheeseburger, a bag of chips—in between treating critically ill patients. And though I'd never been a junk food junkie, I suddenly found myself craving sugary treats like soda or donuts, the effects of which translated to an expanding waistline.

I eventually moved on to become a resident in neurology, working a more reasonable schedule. But had I kept up those habits long enough, I have no doubt that I would have piled on more pounds. Longer still and I would have likely developed elevated blood sugar, high blood pressure, perhaps even diabetes.

They're the same problems I encounter today in many of my patients, who are among the 68 percent of American adults who are overweight or obese. But they almost never come to see me *because* of their weight. Instead, they walk through my door in search of solutions for their memory lapses or foggy thinking. They have no idea that their bulging abdomens are almost certainly shrinking their brains, directly and indirectly. And they don't know that their high-fat, high-sugar diets have changed their brains in ways that make them less likely to make good food choices in the future.

My patients are almost always stunned when I explain the link between excess weight and a shrinking brain. And they're usually eager to get started reversing the damage. I have a program to help them do just that, but first I explain for them exactly how extra pounds—and their food choices—reshape their CogniCity.

Excess Weight, from Bad to Worse

Often when scientists talk about the dangers of being overweight they focus first on the obese. That's understandable. The effects of excess weight tend to be most dramatic in the obese, whose body mass index (or BMI, a measurement of weight as it relates to height) is 30 or above.

But while being obese may be very bad, being merely overweight (with a BMI of 25 to 29.9) isn't good either. As you'll soon read, excess weight takes a toll on memory and cognitive function and also raises the risk of developing pre-diabetes, metabolic syndrome, and type 2 diabetes, all of which carry you further along the brain-shrinking continuum. And while weight doesn't solely determine your likelihood of developing these conditions, generally speaking, the more you weigh the greater your risk. Here's a quick look at each:

Pre-diabetes, or impaired glucose tolerance, occurs when blood sugar is high but not high enough to meet the threshold for diabetes. It is often the first hint of increased glucose intolerance—a sign that your body is becoming unable to regulate glucose levels in the blood—and it affects an estimated seventy-nine million Americans, according to the American Diabetes Association.

Metabolic syndrome is defined by the International Diabetes Federation as a combination of central obesity (abdominal fat) and two or more of the following: a triglyceride level of more than 150 mg/dL; an HDL level of less than 40 mg/dL (or less than 50 mg/dL in women); a blood pressure of 130 over 85 or more or previously diagnosed hypertension; or a fasting glucose level of more than 100 mg/dL or previously diagnosed type 2 diabetes. Metabolic syndrome raises your risk of developing coronary artery disease, stroke, and type 2 diabetes.

Type 2 diabetes is a condition in which the body either doesn't produce enough insulin—the hormone that helps regulate blood sugar levels—or the body's cells have become resistant to insulin. Either way, the result is elevated

The Obesity/Brain Health Continuum

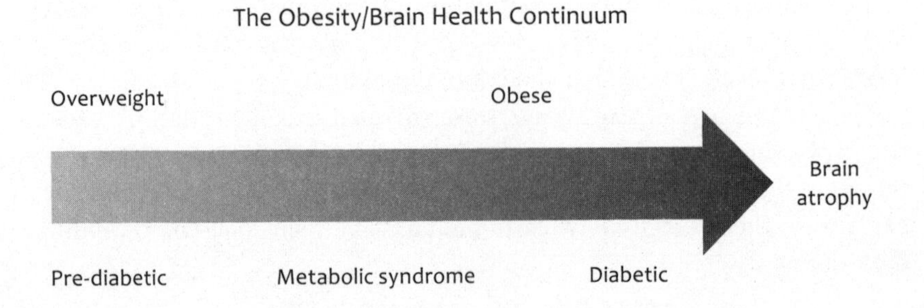

blood sugar levels that cause damage to blood vessels. Type 2 diabetes is usually diagnosed in adults. Some 25.8 million Americans, or 8.3 percent of the population, suffer from the disease today. For simplicity, when I refer to diabetes, you can assume I'm talking about type 2 rather than type 1 diabetes, which is a condition usually diagnosed in childhood and in which the body produces no insulin. Diabetes and excess weight can occur independent of each other: you can be diabetic but not overweight or overweight but not diabetic.

Although it's treatable, diabetes can contribute to the development of a variety of serious health problems, including heart disease, stroke, kidney disease, high blood pressure, nerve damage, and blindness. Each of these conditions contributes to brain atrophy, with effects on brain function that range from subtle to severe.

What's Going On Under the Skull

Just how excess weight, obesity, and diabetes affect the brain is still an area of intense research. But we're gathering more pieces of the puzzle every day.

For starters, we know that obesity is tied to low BDNF, the brain-healing, neuron-fertilizing protein we need for a strong brain. In one study of seventy-three obese children and forty-seven non-obese children between the ages of seven and nine, BDNF tended to be lower in obese children compared to their leaner peers. What's more, after two years those who had lost weight, practiced sports, and had a low-carb diet saw an increase in their levels of BDNF.[1]

In another study, researchers in Denmark measured BDNF levels in obese and non-obese men and women with and without diabetes.[2] People with diabetes had lower levels of BDNF regardless of their weight, suggesting that diabetes may have its own effect on BDNF.

Perhaps the more obvious mechanism of brain shrinkage, though, is the reduced blood flow to the brain that accompanies excess weight, pre-diabetes, metabolic syndrome, and diabetes. Metabolic syndrome, for example, may be associated with a reduction in blood flow to the brain of up to 15 percent.[3] That's an alarmingly high number—remember, neurons need oxygen to stay healthy, as do the helper cells that maintain the brain's highways.

Excess weight also contributes to the risk of developing vascular problems, such as high cholesterol and high blood pressure, and ultimately raises the risk of stroke. Generally speaking, the more excess weight you carry, the greater your risk.

And excess weight contributes to the risk of OSA, another condition of oxygen starvation. Excess pounds also very often (though not always) limit a person's mobility, sometimes drastically. That lack of exertion lowers blood flow to the brain and robs a person's brain of oxygen and nutrients.

In addition, the elevated cortisol levels associated with being overweight or obese may damage the brain. Elevated cortisol not only causes the structural changes you read about in chapter 10 but also increases unhealthy brain wave activity. Patients with obesity and high cortisol tend to have too much fast beta activity in the brain and too little of the alpha activity that's associated with a focused, calm, and alert brain. Conversely, the blood flow problems associated with excess weight, obesity, and diabetes can also result in slower activity in parts of the brain. For example, "dead zones" in the frontal lobes caused by major, small, and micro strokes have low or no activity, leading to sluggish thinking. To go back to our brain orchestra, you can imagine a row of tubas in front pumping out slow, low notes, while elsewhere in the pit dozens of violins briskly hit high notes, thanks to excess cortisol.

Scientists still aren't certain about exactly how diabetes does its damage in the brain, although it's clear that high blood sugar contributes to atherosclerosis, which contributes to vascular dementia (caused by strokes of different sizes). It's likely that other elements are at play, too. Insulin resistance may interfere with the ability to break down amyloid plaques—one of the hallmarks of Alzheimer's disease—in the brain, for example. High blood sugar, meanwhile, may contribute to oxidative stress, which damages cells in the hippocampus. Excess cortisol, high levels of insulin in the blood, and inflammation likely combine to make their mark on brain cells, too.

A Crumbling CogniCity

As you know by now, low BDNF, poor oxygen flow, and unhealthy brain activity harm the brain. Excess weight, obesity, and diabetes, in other words, destroy synapses, wither blood vessels, batter highways, and kill neurons.

The result, of course, is a smaller brain, as researchers at the University of Pittsburgh demonstrated in 2010—8 percent smaller, in the case of the obese people they studied.[4] The study subjects were ninety-four elderly men and women enrolled in the Cardiovascular Health Cognition Study, all of whom showed no outward symptoms of dementia or Alzheimer's disease. Using a novel method to create 3-D maps of brain atrophy, the research team showed that having a BMI greater than 30 carried with it the most dramatic effect: significant decreases in the volume of grey matter. But even being merely overweight (with a BMI of 25 to 29.9), rather than obese, resulted in white matter atrophy.

In obese people, the loss of brain volume appeared to most affect certain parts of the brain: the frontal (attention) and temporal lobes (language), the hippocampus (memory), and the anterior cingulate gyrus (mood), for example. The result of such atrophy is foggy thinking, fatigue, poor memory, and irritability.

The relationship between global brain volume (the size of your brain) and BMI is linear, which means any excess body weight is associated with a slight reduction in brain volume, as one study of 114 individuals forty to sixty-six years of age showed.[5] There's no upper cutoff, so a BMI of 40 is worse than a BMI of 35, which is worse than a BMI of 30.

It's important to note that there is a lower cutoff: being excessively underweight is also associated with a smaller brain. In one study, researchers found that patients with anorexia nervosa had reduced brain volumes.[6] Those who gained weight after recovering from anorexia, however, showed a rebound in their grey and white matter volumes.

Abdominal fat appears particularly suited to shrinking the brain, as Dr. Stephanie Debette and her colleagues at Boston University found when they studied the MRIs of 733 participants enrolled in the Framingham Heart Study.[7]

Initiated in 1948 to study factors that cause cardiovascular disease, the Framingham Heart Study started with about 5,200 volunteers aged thirty to sixty-two living in Framingham, Massachusetts, a town set just a thirty-minute drive to the west of Boston. Over the years, the study has ballooned with so-called cohort studies of the children and grandchildren of the original participants. Researchers have followed the participants and their cohorts closely, tracking their physical health and interviewing them at length about their lifestyles. As you can imagine, given its size and scope, the study has been a gold mine of information for experts examining a host of health problems. In fact, the study has spawned more than 1,200 articles in medical journals.

Dr. Debette's was one such study. Participants were cognitively healthy and relatively young, with an average age of sixty years old, but those who had a higher BMI, a larger waist circumference, greater abdominal fat, and a bigger waist-to-hip ratio were more likely to have lower brain volume. Abdominal fat seemed to show the strongest link with reduced total brain volume, while waist-to-hip ratio was associated with more specific reduced volume in the hippocampus.

By now you know the likely end result of such brain structure changes: lowered cognitive performance and an increased risk of dementia.

Certainly, many studies have tied excess weight to an increased risk of dementia. One tracked 6,583 Northern California Kaiser Permanente patients between 1964 and 2006 and found central obesity was associated with a higher risk of dementia. Those who were overweight or obese but didn't have central obesity had an 80 percent increased risk of developing dementia thirty years later, compared to non-overweight people. Those who had central obesity, however, were 3.6 times more likely than non-overweight peers to develop dementia—that's a 360 percent increased risk. Even being merely overweight was associated with a 234 percent increased risk if the person had belly fat.[8]

Of course, this doesn't mean that every overweight person will become demented or that every obese person is less smart than every thin person. Remember, we all have varying cognitive abilities to begin with, and there are other factors at play. If you're an obese college professor who plays the violin, exercises, and belongs to a bird-watching club, you may be better off than a couch potato who weighs less but never exercises or explores his world. Still, it does mean that the brains of overweight people aren't in the best possible shape and are more susceptible to decline later in life.

Perhaps an even bigger brain shrinker than obesity is diabetes, which has been tied to a smaller hippocampus[9] and greater risk of stroke. One study found that having diabetes doubled the risk of stroke in the elderly and that young people with diabetes increased their risk of developing stroke by nearly six times.[10]

And while many of the worst effects of diabetes make their biggest mark as we age, they are by no means limited to the elderly. Just ask researchers at the New York University School of Medicine, who studied obese adolescents with type 2 diabetes and found that they had lower scores on tests of intellectual functioning, verbal memory, and psychomotor efficiency.[11] The study, published in 2010, compared obese teens with type 2 diabetes to teens who were obese but didn't have the disease, which points to diabetes having a greater effect on the brain than mere obesity.

In 2011, the team followed up on those findings with a report that obese adolescents with type 2 diabetes had measurably smaller hippocampi and frontal lobes than obese teens without diabetes.[12] Those results were rather shocking. At a time when their brains are supposed to be developing, diabetic teens are actually experiencing brain *shrinkage*. And the number of such teens is on the rise. Imagine their future. If left untreated, their diabetes will continue to damage their brains for decades to come. Looking sixty years down the road, it's not hard to imagine that they'll suffer the consequences, with a higher risk of cognitive decline and Alzheimer's disease.

As a neurologist, I see many different manifestations of diabetes, from the neuropathy that results in a loss of sensation in the toes, to the nephropathy that causes kidney failure, to the retinopathy that robs diabetic patients of their sight. And, of course, I often see the effects of diabetes on the brain. We don't call it "brainopathy," but perhaps we should. The impact is as life altering—if not more so—as any damages below the neck.

How High Cholesterol Factors In

High blood cholesterol is a vascular risk factor that often goes hand in hand with excess weight. And since it's one of the conditions that slows blood flow to the brain, blood cholesterol bears mentioning in any discussion of brain shrinkers.

If you're a bit rusty on your cholesterol facts, here's a quick refresher. High-density lipoprotein (HDL) is often referred to as "good cholesterol" because it helps to remove excess cholesterol from your blood. Low-density lipoprotein (LDL), on the other hand, is considered "bad cholesterol" because it can build up on the walls of your arteries. Triglycerides, a type of fat, are also part of the overall picture, which we call your "cholesterol profile." Ideally, I like to see my patients keep their HDL above 60 mg/dL and their LDL and triglycerides each below 100 mg/dL.

An unhealthy cholesterol profile can lead to atherosclerosis, a buildup of cholesterol plaques in the blood vessels, including those in the brain, which sixty years ago was referred to as "hardening of the arteries." The name is fitting: if you were to touch the blood vessels affected by atherosclerosis, you'd actually find them to feel crunchy under your fingertips.

High cholesterol doesn't directly produce symptoms, so it can go undetected. If unchecked, the plaques associated with high cholesterol trigger inflammation, which causes further growth of atherosclerotic plaques. The resulting high blood pressure (you'll read more about this in chapter 12) adds to the damage by putting continuous pressure on the artery walls, weakening them, and increasing the likelihood of plaque buildup. Over time the interior channel of the blood vessel—called the lumen—narrows, slowing the flow of blood. Eventually the blood vessel can be blocked entirely, cutting off blood supply.

Atherosclerosis can affect blood vessels in any part of the body. When it occurs in blood vessels leading to the heart, it can cause a heart attack. When it occurs in the brain or in the arteries leading to the brain, the result is a stroke. Whether they're large strokes, small strokes, or even micro strokes, such "brain attacks," as you'll read in chapter 12, starve the brain of nutrients and kill neurons. The result can be a serious loss of function—physical, cognitive, and emotional.

High levels of LDL, and low levels of HDL, have also been tied to cognitive decline, independent of obvious signs of stroke. One study, led by my friend and colleague University of California professor Dr. Kristine Yaffe, whom I consider a star in the field of brain health research, found that older women with the highest levels of LDL were more likely than those with low LDL to have mild cognitive impairment.[13]

High LDL cholesterol, and even borderline high LDL cholesterol, especially in combination with high blood pressure, also increases the risk of developing Alzheimer's disease later in life, as one study of nearly 1,500 people in Finland found.[14]

Even more compelling, perhaps, is evidence that ties low HDL to poor memory—even in midlife.[15] Researchers made that link by combing through data on 3,763 participants of a long-term study of more than 10,000 British civil servants. Study participants with low HDL (less than 40 mg/dL) were 27 percent more likely to show memory deficits by age fifty-five than those with high HDL (60 mg/dL or more). That number rose to 53 percent by the age of sixty. Knowing this often helps people get motivated to improve their cholesterol profile.

More research needs to be done to determine just how high HDL affects brain function, but it's important to know that both HDL and LDL matter when considering brain health. Even more important, such effects can begin to take a toll—robbing you of brain reserve—long before you reach old age.

Obesity, smoking, poor diet, lack of exercise, diabetes, and other factors can all boost your risk of developing high cholesterol. Heredity can also play a role.

By now you're probably ready for some good news. Fortunately, the brain-draining effects of high LDL seem to be reversible. In the University of California study by Dr. Yaffe and her colleagues, for example, women who lowered their levels of LDL over a four-year period had better cognitive function than those whose LDL levels remained high.

The study also noted a positive correlation between statin use and better cognitive function, independent of whether or not study subjects taking them actually lowered their LDL levels. Statins are cholesterol-lowering medications that are highly effective in reducing the risk of heart attack.

Why might statins offer neuroprotective benefits? Some experts suggest that in addition to their LDL-lowering properties, statins may act as an anti-inflammatory agent.

Other studies have shown a reduced occurrence of dementia in people who take statins, although they don't yet prove that taking statins reduces a person's risk of dementia. (We don't know, for example, whether people who take statins happen to also do other things that result in them being less likely to develop dementia.) And there's no proof that statins can slow the progression of Alzheimer's disease.[16]

Interestingly, although some people have complained of cognitive problems after they started taking statins, studies show this is actually a rare occurrence and that the benefits of statins far outweigh the small risk of minor cognitive problems.[17]

How Your Brain Affects Your Weight

In the winter of 2012, a fifty-year-old man named Randall walked into my office. Or rather, he wheeled in. At 450 pounds Randall could walk, but just barely. Instead, he spent most of his day in a motorized wheelchair.

Randall hadn't always been obese, but marital strife a decade earlier had sent Randall into a tailspin of overeating. "I was so stressed and sad that I just ate uncontrollably," Randall told me on his first office visit. Over the years, Randall's weight had ballooned from 200 pounds to 290. By then his knees had begun to ache. Eventually, as he passed the 300-pound mark and kept going, his knee pain would essentially cripple him.

Randall was a smart software engineer in New York, making a nice living creating computer networks for business clients. He was skilled in his field, but as Randall's health went downhill, something else happened: Randall's business began to suffer. By the time he came to see me—worried that he might be developing Alzheimer's disease—Randall had had two major projects fail, had stopped working, and had started collecting disability payments. Once vibrant and gregarious, Randall was now flat-out miserable.

After my usual complete evaluation, I told Randall that he didn't have Alzheimer's disease and that there was plenty we could do to help him regain his memory, clarity, and creativity. But Randall would have to change his body. And to do that, he'd have to change his brain.

To understand why, you first need to know the role the brain plays in the development of excess weight and obesity. While there are many complex psychological underpinnings to obesity, there's also some very clear neuroscience involved. That science starts below the neck with three hormones—insulin, leptin, and cholecystokinin (CCK)—that are released every time we eat. These hormones act as a negative feedback system, similar to a thermostat, helping you to stop eating when your glucose levels are high (insulin), your fat levels accumulate (leptin), or when your stomach is full (CCK). Unfortunately, this thermostat stops working when a patient becomes obese.

Of course, there are parallel pathways involved as well. Researchers in Finland, for example, presented evidence in 2012 that the brain's reward system in response to food is more active in obese people.[18] In that study, obese people shown pictures of food had increased activity in the brain's reward area and decreased activity in the prefrontal cortex, the main part of the brain's braking system, which helps prevent overeating.

It's a terrible combination of factors. And it may explain (along with the faulty "thermostat") why many people find it so hard to stop overeating. As you gain weight, changes in your brain's reward system may make you crave certain foods. At the same time, your brain's braking system is asleep at the

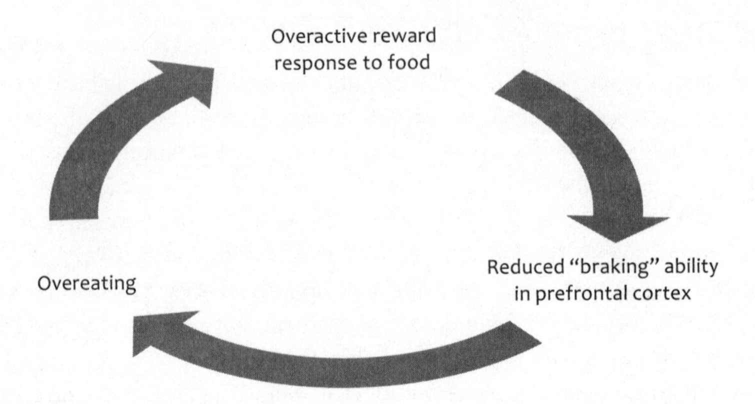

wheel, failing to help you control your impulse to seek out the short-term reward. This is why I sometimes refer to obesity as a prefrontal cortex problem more than a food problem. It's the brain that's driving the choices you make, after all, telling you to choose french fries when you really should order a side salad.

Treatment: Barreling Back from the Brink

When I told an ecstatic Randall that many of his memory and cognitive problems could be reversed, I wasn't exaggerating. As severe as his weight problems were, there's ample evidence that weight loss and the treatment of diabetes have a dramatic effect on brain health.

One study of bariatric surgery patients, for example, found that those who'd lost weight after surgery were able to improve their cognitive testing scores and reverse damage to their brains' highways in just twelve weeks. For that study, researchers tracked the progress of 109 bariatric surgery patients enrolled in the Longitudinal Assessment of Bariatric Surgery project, plus 41 obese control subjects who hadn't had bariatric surgery.[19] Study subjects were given a cognitive evaluation at the start of the study and again twelve weeks later.

At the start, about one-quarter of the bariatric surgery candidates studied had below average cognitive performance—so much so that their scores placed them in the "impaired" range, chiefly due to deficits in memory and reasoning abilities.

But three months after surgery, patients who'd lost the most weight substantially improved their memory performance. (Those who didn't lose weight, meanwhile, saw their scores decline twelve weeks later.)

Even when the weight loss itself is less dramatic, the results can be long lasting. Researchers in one diabetes prevention program found that overweight people who had modest weight loss—about fourteen pounds—were able to drastically reduce their likelihood of having diabetes. Even better, the benefits of such weight loss can persist for ten years, even if the person regained some of the weight during that time.

Randall's many interlocking health problems—OSA, depression, high cholesterol, and high blood pressure—actually offered us many avenues for improvement. And every small step toward better physical health would be one step away from further cognitive decline.

But *really* growing Randall's brain would require him to do something massive. Randall would have to lose a tremendous amount of weight. Now, this is no small challenge, as Randall knew. He didn't get to 450 pounds overnight, after all. Instead, his extreme condition had developed over years, starting with some excess weight and then snowballing, with the ever-mounting toll of an interconnected set of conditions. As he piled on the pounds, for example, he developed obstructive sleep apnea, which contributed to the development of depression. The effects of OSA and depression left him fatigued and unmotivated, slowing his activity level dramatically. With less activity, Randall gained yet more weight. Soon, that extra weight triggered knee pain that kept him off his feet entirely. High cholesterol and high blood pressure, meanwhile, slowed the flow of blood to his brain, setting off silent strokes that subtly hampered his brain function.

Over time, Randall would work with my staff to improve his overall health by increasing his exercise, changing his diet, lowering his blood pressure and cholesterol, and treating his depression and sleep apnea. Exercise would help to not only grow his brain but also reduce his insulin resistance. A high-protein, low-carb, brain-healthy diet, as I described in chapter 5, would also help.

Changing his body would help Randall grow his brain. But he would also have to change his brain in one key way: he would have to override his brain's faulty reward system again and again, until it reset itself and began functioning as it had before he'd become overweight. Eventually, his prefrontal cortex would be back in action—so much so that making healthy food choices would actually be easier. That, of course, would help him change his body.

All of it would rely on neuroplasticity—the brain's ability to grow and change—which, as you now know, we all possess. Even Randall, whose downward spiral was so pronounced, could change his brain and reset his course on a path that would send him instead spiraling *up* to a bigger, better brain.

Of course, it all starts with one key step: recognizing that your approach to eating and exercise needs a serious overhaul. Randall had his aha moment sitting in his wheelchair in my office, with his MRI results on the computer screen before us. Once he connected his cognitive symptoms with his health problems, he was energized like never before. Always a thinker and a planner, he quickly came up with a concrete path forward: immediate dietary changes followed by knee surgery, then exercise and enough weight loss to allow him to qualify for bariatric surgery. After surgery, he would continue to work on his diet and exercise, helping him lose weight and maintain those results.

Randall followed through on his plan and after his bariatric surgery, he continued to work on his diet and exercise, dropping down to just over 200 pounds. With the weight off and his health vastly improved, Randall reported feeling mentally stronger and clearer than ever. Best of all, his fear of Alzheimer's had become a distant memory.

I often have my patients sit down and write up a specific action plan. In fact, it's part of their brain fitness program passport. When they do, I always remind them to be realistic, shooting for one to two pounds of weight loss a week rather than rapidly dropping weight. And I advise them to spend time seeking out a weight-loss solution that meets their unique needs. I've had patients who have loved the social support they got by joining groups like Overeaters Anonymous or any one of the many group-oriented weight-loss programs. Others have found online, individualized tools work best, while still others have found certain cookbooks or diet books to guide them. As long as it promotes responsible weight loss, the method is less important than the end goal: upending that downward spiral and riding the result to a bigger brain.

CHAPTER TWELVE

Brain Attacks, Large and Small

IF YOU'D SEEN him rattling off figures in the boardroom or locking up his latest multimillion-dollar deal, you'd never have suspected that my patient, sixty-two-year-old Marc, was about to suffer a stroke. Vibrant and energetic, he looked anything but unhealthy. Looks can be deceiving, though.

Marc, who lived in Philadelphia and traveled frequently, had a high-stress lifestyle, plus high cholesterol, high blood pressure, and a years-long habit of getting by on three or four hours of sleep a night. Marc admitted he had a classic type A personality. In recent years, he'd been feeling increasingly tired. Then, walking out of a meeting one afternoon, he noticed that his left hand was clumsy and his walking unsteady. He chalked up the strange sensations to too many hours in the office and decided to leave early that day. Hours later, with his hand still feeling clumsy, he finally called his primary care physician, who advised him to head to the emergency room immediately. Once there, Marc got some bad news. An MRI of his brain showed an apple-size stroke in his cerebellum—the brain region important for coordination and balance.

Marc was stunned to find himself among the roughly eight hundred thousand Americans a year who suffer a stroke. Like many, he'd always viewed a stroke as a tragedy that befalls only the very ill or elderly.

As upsetting as the news was to him, it was about to get worse. When he came to see me, I discovered that Marc's blood pressure was a shocking 170 over 110 and his pulse was 100. He also had borderline sleep apnea, borderline diabetes, and a memory performance that was below average for his age and education level. I explained to him that his stroke was just the tip of the iceberg. The conditions that had led to it, after all, were still quietly at work. Not only could they work to shrink his brain on their own, but they were also no doubt producing microscopic strokes too small to be seen on an MRI. And

The Stroke/Brain Health Continuum

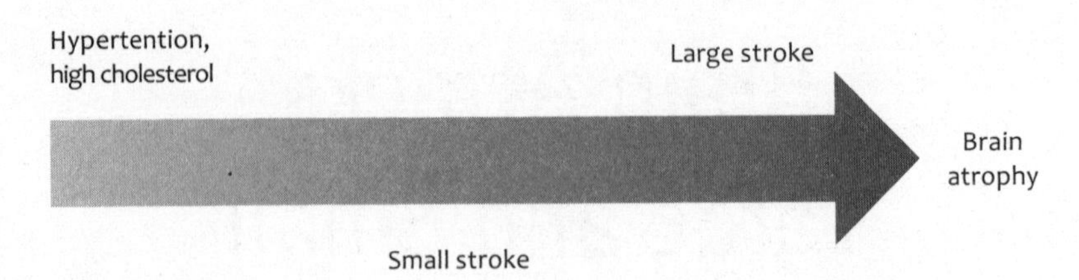

if his underlying health problems remained untreated, they could lead to more small strokes or even a major stroke within months.

In fact, just as the brain-shrinking effects of obesity (or sleep, or stress) progress along a continuum, so too do the effects of restricted blood flow to the brain, with vascular risk factors like high cholesterol and hypertension on one end, small strokes in the middle, and a major stroke on the other end.

Marc, who'd always been "too busy" to worry about reducing his stress, lowering his blood pressure and cholesterol, exercising, and getting adequate sleep, would have to change his life to save his brain. He could do it, but first he'd have to understand why he developed a stroke and how strokes can be prevented.

Higher Pressure, Smaller Brain

Marc's stroke happened for a reason. It occurred because blood flow in his cerebellum was blocked.

His high cholesterol played a part in creating that blockage. As you'll probably recall from my discussion of excess weight and obesity, high cholesterol causes plaques that can build up and narrow blood vessels. Those narrowed vessels slowed the flow of blood, and thus oxygen, within Marc's brain. But they also raised his blood pressure by forcing his heart to pump harder to push blood through. Hypertension then worsened the already bad situation in a plaque-filled blood vessel, helping to thicken and stiffen the blood vessel even more.

Like high cholesterol, hypertension is a major risk factor for stroke and brain atrophy. But even in the absence of a major stroke, the vascular damage caused by high cholesterol and hypertension result in lower blood flow to the brain, shrinking synapses, damaging brain highways, and killing cells.

The result is a smaller brain, as researchers discovered at the Institute for Ageing and Health at the University of Newcastle upon Tyne in the United

Hypertension Risk Factors

Hypertension affects sixty-eight million American adults, according to the Centers for Disease Control. It typically produces no obvious symptoms and thus often goes undiagnosed. Risk factors for hypertension are:

- excess weight,
- a high-sodium diet,
- diabetes,
- alcohol abuse,
- inactivity,
- smoking, and/or
- obstructive sleep apnea.

Hereditary factors can be at play too.

Kingdom. Those researchers examined brain MRIs of 103 people with hypertension and 51 without hypertension.[1] They found that even moderate hypertension resulted in smaller overall brain volume, a smaller hippocampus, and increased white matter damage.

Hypertension in midlife has also been linked to an increased risk of cognitive decline in late life.[2] In one study, researchers first examined study subjects while they were in midlife and then checked back twenty-five years later. Those who'd had untreated hypertension in midlife were 2.6-fold more likely to have Alzheimer's disease later in life. Researchers have also found that earlier hypertension is tied to smaller overall brain volume and evidence of numerous silent strokes later in life.

A Concern for All Ages?

There's growing evidence that the brain-shrinking effects of hypertension aren't just a worry for late life. In fact, researchers are beginning to uncover proof that the cognitive effects of hypertension may be felt far sooner than we once thought.

One such expert is my colleague Dr. Charles DeCarli, a stroke specialist who is a professor and director of the Alzheimer's Disease Center at the University of California, Davis. Dr. DeCarli has long studied the effects of vascular risk factors on the aging brain. In a 2011 study published in the journal *Neurology,* Dr. DeCarli and his team reported their findings regarding 1,352 participants in the Framingham Heart Study, who were in their midfifties when first assessed for vascular risk factors and had no signs of dementia, stroke, or other neurological disorders.[3]

Looking at brain scans and cognitive testing performed ten years after the subjects had first been enrolled, the research team found that people who'd had hypertension in midlife were more likely to show signs of white matter damage on their brain MRIs and worsening executive function on cognitive testing. It was new proof that the harm from midlife hypertension can be seen in subjects who were far from old.

In fact, hypertension likely starts to make its mark in the brain even before midlife, says Dr. DeCarli, who plans to do further research to find out just when that happens. It's especially important given the growing number of obese and hypertensive children who, if allowed to remain so, may enter midlife with a considerable load of damage marring their brains.

Stroke: A Brain Killer

Chances are you hear rather regularly about strokes. We're constantly reminded, especially as we age, that strokes kill. After all, stroke is the third leading cause of adult death in the United States (behind only heart disease and cancer) and *the* leading cause of disability in adults.

But if, like Marc, you've never paid attention to the details of stroke, now is a good time for a refresher course.

A stroke is the death of brain cells that occurs when blockages in the arteries in the brain or bleeding in the brain prevent blood from flowing to the area. Without the oxygen and nutrients blood brings, the affected neurons die.

A "hemorrhagic" stroke occurs when a blood vessel ruptures—as can happen with very high blood pressure or a sudden spike in hypertension—cutting off blood flow to a part of the brain. In a minority of cases, an irregular heartbeat—called atrial fibrillation—causes the formation of blood clots in the heart, which then travel to the brain, block blood vessels, and cause "embolic" strokes. But most strokes are "ischemic" strokes, which occur due to vascular disease that thickens, stiffens, narrows, and eventually clogs a person's blood vessels.

Blood flow blockage in large blood vessels results in "major" strokes, which cause sudden, severe symptoms, such as the loss of feeling on one side of the body or the loss of the ability to speak. Just where in the brain the stroke strikes determines which functions are lost. If the damage occurs in the visual cortex at the back of the brain, for example, a patient may lose vision. If it's in the motor cortex, toward the middle of the brain, a person may lose the ability to move an arm or leg on one side of the body. In rare cases, the hippocampus or temporal lobes are affected, resulting in acute memory loss and aphasia, or loss of speech. Often stroke affects mood, motivation, personality, and speed of cognitive processing too.

Ischemic strokes can also occur in the small blood vessels in the brain. Instead of killing a large swath of brain tissue, such strokes lead to the death of a small patch of cells about the size of a grain of rice. Such small strokes are often dubbed "silent" strokes because they typically don't cause obvious symptoms when they occur. But in reality, they are anything but silent; thousands of silent strokes that take place over decades can create thousands of small dead zones. This results in not only slowed thinking but also depression and a lack of energy.

Eventually, strokes of all sizes can add up to marked atrophy in the whole brain, resulting in vascular dementia. In one study, for example, people who'd had strokes were twice as likely to develop dementia later in life.[4]

In recent years we've discovered the existence of yet smaller strokes, so tiny they can only be seen under a microscope. While these are too small to be detected on MRI, they are far from benign. As my friends and colleagues, Johns Hopkins University professors Juan Troncoso and Richard O'Brien, have shown, people who have such microscopic strokes are at a high risk for developing late-life dementia.[5]

Signs of Stroke

If Marc had experienced chest pains instead of clumsiness in his hand and trouble walking, chances are he would have sought help immediately. People tend to take heart attacks seriously. But many fail to recognize the signs of a stroke. Knowing that, I always tell my patients to think of a stroke as a "brain attack" and treat any sudden neurological symptoms just as seriously as they would if they thought their hearts were in crisis.

Such caution is all the more important because a quick medical response to a stroke can make a huge difference in the outcome. Sufferers of an ischemic stroke who are given the clot-busting drug tPA (tissue plasminogen activator) within three—and possibly as late as four and a half—hours of a stroke, for example, have a far better chance of reducing long-term damage to their brains than those who don't get treatment.

Often, however, people don't recognize the symptoms, put off seeking help, or even try to "sleep it off." Just 3 percent of patients suffering an acute ischemic stroke actually receive a clot-busting drug in time, according to the American Stroke Association. Marc is a perfect example of someone who didn't seek immediate attention for a stroke.

Since getting quick help can be so beneficial, it's critical to know the signs of a stroke, and to heed them. They include a sudden onset of:

- numbness or weakness in the face or extremities, especially on one side of the body;
- unexplained confusion;
- difficulty speaking or understanding speech;
- vision loss in one or both eyes;
- difficulty walking;
- dizziness or loss of balance; and/or
- a severe unexplained headache.

The National Stroke Association proposes using the mnemonic FAST for quickly evaluating someone with a suspected stroke:[6]

F for *face:* Ask the person to smile. Does one side of the face droop?

A for *arms:* Ask the person to raise both arms. Does one arm drift downward?

S for *speech:* Ask the person to repeat a simple phrase. Is the person's speech slurred or strange?

T for *time:* Don't waste time! If you observe any of these signs, call 911 immediately.

Wiping Out Neighborhoods and Highways

When you look at the MRI of a stroke victim the devastation is obvious: those patchy spots dotting the brain represent dead brain cells. You don't have to be a neurologist to realize the result is a loss of function.

But in addition to the brain cells in grey matter killed by a stroke, limited blood flow can result in damage to white matter. In CogniCity, you can think of it like this: vascular disease can cause a stroke, which decimates the neighborhood, but it can also erode the highways and roads that connect those neighborhoods. Remember, those highways are made of delicate live cells. They need maintenance in the form of a constant supply of oxygen and nutrients.

In the past, many people—even doctors—have dismissed such damage to fiber bundles in the brain as the inevitable result of aging. In fact, radiologists often note it in their reports as "non-specific white matter changes associated with aging."

But white matter damage isn't inevitable and it is by no means benign. In recent years we've gotten a clearer picture of just how white matter damage affects the brain. In one study, for example, Dr. DeCarli and his colleagues found that white matter damage even in healthy people is associated with lower brain volume and poorer performance on cognitive tests.[7]

For anatomical reasons, ischemic damage just happens to be more common in the front of the brain than in the back of the brain. Damage that occurs here results in declines in memory, attention, and the ability to perform complex cognitive tasks. Patients with ischemic injury in their frontal lobes may be flummoxed by the "easy check-in" kiosk at the airport or by doing their taxes. "They're all things that older people get into trouble with," notes Dr. DeCarli. "And anything that accelerates that then accelerates cognitive aging."

Some amount of damage in the brain can be overcome, but multiple types of damage are likely to lead to serious irreversible problems. If you have a few small strokes in your frontal lobes, for example, but your temporal lobes are in good shape, you might find that complex mental tasks are challenging but you can still get by in life. But if both your frontal and temporal lobes are damaged, your cognitive decline will almost certainly affect your ability to function independently.

Strokes, meanwhile, can also hamper healthy brain activity. That makes sense since swaths of the brain that die in a stroke don't function. In fact, in the days before high-tech imaging was widely available, strokes were often detected using EEG, which could easily pick up dead zones. The lack of normal brain activity associated with a stroke explains some of the behavior we see in people with major strokes or extensive damage from silent strokes. Take the example of the stereotypical elderly man whose frontal lobes have been destroyed by a stroke. There's a reason he's cranky and erratic: the section of

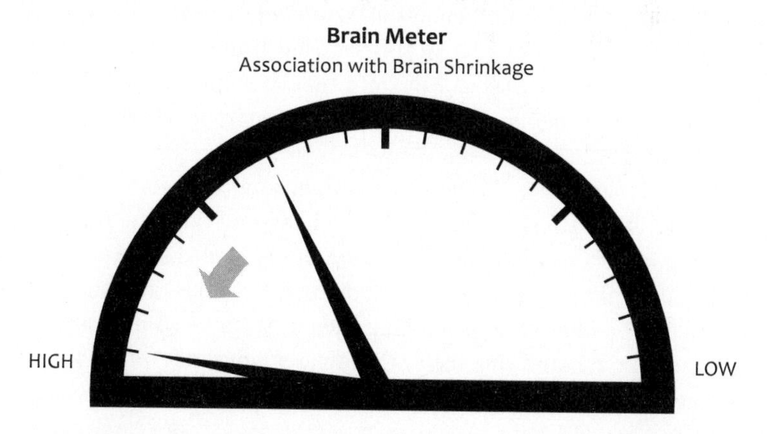

Multiple microscopic or silent strokes are associated with brain atrophy as well as damage to the brain's highways, as are major strokes. The more (or more severe) the strokes, the worse the brain shrinkage.

his brain responsible for impulse control has been silenced. If you hooked him up to EEG, you'd find his frontal lobes to be barely buzzing.

Warning: Stroke Ahead!

When you consider what's likely going on in the brain of someone who has had an ischemic stroke it's no surprise that he or she is at greater risk of a future stroke than someone who hasn't. It only takes one blocked artery to cause a stroke, but the conditions that led to that blocked artery are almost certain to be having a similar impact in other arteries. This means conditions are likely in place for future strokes, large or small.

In addition, transient ischemic attacks (TIAs)—or ministrokes—are a warning sign of a future stroke. TIAs are caused by the temporary blockage of a blood vessel, which fortunately spontaneously resolves. The symptoms—such as paralysis (usually on one side of the body), loss of balance, or slurred speech—dissipate within one hour and typically cause no lasting damage. Still, a TIA may be a sign of things to come; roughly one-third of people who have TIAs experience a stroke within a year.

Who's at Risk?

If you're like Marc, you might have thought you had a few decades more before you had to worry about strokes. But a stroke can strike well before old age, and there's evidence that strokes may now be affecting young people more than ever before.

A 2012 study published in the journal *Neurology,* for example, showed that the average age of first-time stroke sufferers fell from seventy-one years old in 1993 to sixty-nine in 2005.[8] Not only that, but the under-fifty-five age group constituted 19 percent of total stroke sufferers in 2005, compared to just 13 percent in 1993.

For all ages, there are a variety of conditions and behaviors—called vascular risk factors—that affect blood flow in the body and raise the risk of stroke. They include hypertension, diabetes, carotid artery stenosis, peripheral artery disease, atrial fibrillation, coronary heart disease, sickle cell anemia, high cholesterol, inactivity, obesity, and smoking. Heredity may also play a role.

The more risk factors you have, the greater your risk of stroke. If you want a rough idea of your risk, you can use the National Stroke Association's "Stroke Risk Scorecard."[9]

We're also increasingly learning about other factors that may raise the risk of having a stroke:

Stress and Type A Personality In one study, researchers found that people who had experienced a major negative life event in the prior six months were nearly four times more likely to suffer a stroke than those who hadn't.[10] For the study, researchers compared 150 middle-age adults in the stroke unit of a hospital with 300 non-stroke sufferers in the same community.

In addition to measuring how much stress study subjects had been under, the research team assessed their behavioral patterns, looking for type A personality traits, such as hostility, aggression, and a quickness to anger. As it turned out, having a type A personality also was linked to stroke, doubling the risk.

Poor Diet Unhealthy eating that leads to obesity may increase the risk of stroke by virtue of boosting BMI. But high consumption of sodas and a high-sugar, high-salt diet ups your risk of stroke, too. One study, using data from more than 127,000 people who were tracked over more than twenty years, found that having greater than one serving of soda—even diet soda—a day increased the risk of stroke by 16 percent.[11]

Sleep Disorders Obstructive sleep apnea and insomnia raise the risk of stroke, as you read in chapter 9. Not only that, but people who have a stroke and obstructive sleep apnea are 76 percent more likely to die following the stroke, compared to those who have a stroke but don't have sleep apnea.[12]

Fighting Stroke

The heartbreaking truth about strokes is that once brain cells die, they don't come back. But even so, the brain has an innate ability to repair itself and so there is tremendous healing activity that takes place after a stroke. Surviving neurons regroup and rewire, with disparate brain areas chipping in, in an effort to compensate for neurons lost to a stroke. Often, stroke sufferers can regain function over time.

I feel passionate about making sure none of my patients experience a stroke. Along with the staff of my brain fitness program, I help patients appreciate the fact that up to 80 percent of strokes can be prevented, and I work with them to understand why and how. Much of the prevention effort comes down to reducing vascular risk factors, such as hypertension, high cholesterol, diabetes, obesity, and a sedentary lifestyle. Doing so may also help reverse some of the damage that is done, even before a large stroke occurs.

At my Brain Center, stroke prevention efforts focus on increasing blood flow to the brain, primarily through improved cardiovascular fitness. Exercise has been shown to aid not only in stroke prevention but also in stroke recovery. In one study, six months of exercise after a stroke resulted in a 50 percent improvement in memory and overall brain function.[13]

To determine patients' fitness levels we measure VO_2 max by having them exercise strenuously on a stationary bike. They then meet with our exercise physiologist and start a personalized program of diet and exercise to boost their fitness by 5 percent every five weeks. We also teach them about heart-healthy food, as well as foods that are particularly effective at reducing the risk of stroke and boosting brain growth (see chapter 5).

One such food is chocolate, which was shown to reduce the risk of stroke in a ten-year study of Swedish men. In that study, researchers found that when they adjusted for all other factors, consuming a third of a cup of chocolate—or 63 grams per week—reduced the risk of stroke by about 17 percent.[14]

Remember Marc, my stroke patient? Before I'd explained to him all that you've just read, I'm not sure he really would have committed to changing his life. After all, he had put off exercising—and even sleep!—for years, with the lament "I just don't have time!" I'm sympathetic. Everyone, myself included, feels short on time at some point in his or her life.

After an hour with me, though, Marc understood that he was on a path toward increasingly noticeable cognitive difficulty at best, and a massive stroke or death at worst. Suddenly, finding time seemed the least of his problems.

Marc cut back on work, started weekly meditation sessions, and shifted his outlook from one that screamed "live to work" to one that calmly (and happily) declared "work to live." He lowered his cholesterol and his blood pressure. With vigorous exercise, his exercise capacity (measured by VO_2 max) increased by 15 percent in three months. Since his sleep apnea was merely borderline, we opted to treat it with weight loss, which was successful.

True to form, Marc tackled his brain fitness efforts just as he would have approached any venture: with determination and grit. Within months he was fit, strong, and feeling mentally sharper. He had reduced his future risk of stroke and no doubt had grown his brain in the process.

CHAPTER THIRTEEN

The Battered Brain

BY ALL ACCOUNTS, my patient Gary had some spectacularly bad luck throughout his life. It started with a particularly gruesome accident when he was just a child. Riding in the back of a pickup truck, he not only fell out but also somehow became trapped and was dragged behind the truck. Among his injuries was a fractured skull, a traumatic brain injury (TBI) that left him unconscious for four days. Fortunately, he went on to make what seemed to be a full recovery over the next five years.

As an adult, though, he was in for more bad luck. Mouthing off at the wrong moment, Gary, a mechanic from Delaware, found himself in a fight outside a bar one night. The scuffle ended with him taking a hard hit to the head from a baseball bat, an injury that left him blind in one eye. A decade later, he was hit by a truck while walking along a busy road at night. Then, in his sixties, he contracted pneumonia and became critically ill. During his illness, he experienced serious cognitive problems; at his worst, he was so confused that he failed to recognize even his son. Thankfully, Gary recovered, yet again. But many of his cognitive deficits persisted.

When he came to see me, Gary, then sixty-nine, was experiencing profound memory and thinking problems. His son, who accompanied him on his visit, told me Gary had never truly bounced back from pneumonia. "I think he's about 30 percent of what he used to be," Gary's son said, shaking his head.

And no wonder: Gary's MRI showed brain atrophy and extensive white matter damage. His EEG brain map, meanwhile, painted a picture of sluggish brain activity, with excess slow theta waves in his frontal lobes. And Gary's cognitive testing results reflected his brain's poor function: his memory and other cognitive skills were far below average for his age and education level.

His pneumonia had actually been the final straw for a brain that was no doubt low on reserve, thanks to the battering it had taken over the years, starting with that first fall off a truck. In fact, the damage done by that injury no doubt contributed to his likelihood of encountering his future injuries. Reduced function in his prefrontal cortex—which you'll remember is the brain's braking system—was likely a factor in Gary's behavior outside the bar that night, which earned him a crack across the head, further injuring his brain. *That* TBI may have further affected his future ability to make smart decisions, like whether or not to walk alongside a busy road at night, setting him up for his third injury.

Gary's story was fairly dramatic, I'll admit. But every week I see patients who have similar, albeit less extreme, effects from concussions and serious TBIs. Fortunately, neuroscientists have learned a great deal in the past decade about just what happens in the brain when it's injured by trauma, information that has vastly advanced the treatment of concussions and serious TBIs. One of the more important discoveries was proof that, while a major TBI is clearly terribly damaging to the brain, small concussions can cause lasting injury, and multiple small injuries can add up over time.

What Is a TBI?

The term "traumatic brain injury" sounds like something you'd only hear in a hospital emergency room, but TBI is actually becoming an increasingly familiar term in discussions about sports, car accidents, and battlefield injuries. An umbrella term that includes injuries ranging in severity, a TBI is any type of brain injury that results from a blow to the head or a severe shake or jolt that damages the brain. It can be a closed injury, in which there is no penetration into the skull (as you might have if you fell off a motorcycle), or a penetrating injury, in which an object pierces the skull (as would happen with a gunshot to the head).

Each year an estimated 1.7 million people in the United States suffer from a TBI of some sort, according to the Centers for Disease Control. (Some estimates are even higher.) Thankfully, the vast majority of sufferers—about 75 percent—have a mild form of TBI: a concussion. The term "concussion" is often used interchangeably with the term "mild TBI" (and I'll do the same here).

The TBI/Brain Health Continuum

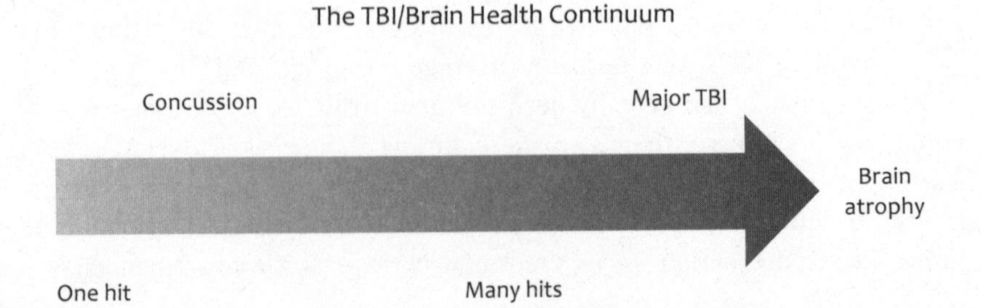

Immediate signs of a concussion, according to the American Academy of Neurology, may include:

- a vacant stare,
- delayed verbal and motor responses,
- confusion or an inability to focus attention,
- disorientation,
- slurred or incoherent speech,
- stumbling or an inability to walk a straight line,
- emotions out of proportion with circumstances, and/or
- memory deficits.

Symptoms, often called "post-concussive syndrome," can appear immediately or within days of the injury. They include:

- headaches,
- changes in vision (including blurring, "seeing stars," or double vision),
- short-term memory loss,
- difficulty concentrating,
- nausea,
- vomiting,
- dizziness or balance problems,
- ringing in the ears,
- sensitivity to light or sound,
- changes in taste or smell,
- mood swings,
- sleep changes (too much or too little),
- depression, and/or
- fatigue.

Symptoms of more serious TBIs can appear immediately or within days of the injury and include:

- confusion,
- coordination problems,
- agitation or aggression,
- difficulty with language,
- weakness or numbness in the extremities,
- loss of bladder or bowel control,
- severe headache,

- repeated vomiting,
- dilation of one or both pupils,
- clear fluids draining from the nose or ears,
- convulsions or seizures, and/or
- loss of consciousness.

For most people, a mild concussion will resolve in a week to ten days with minimal treatment, which might include cognitive and physical rest, plus pain killers to treat headaches. But concussions can be deceptive: seemingly small injuries can produce lasting deficits.

What Happens in the Brain

If you had looked inside Gary's brain after any one of his TBIs you would have witnessed a devastating cascade of events unfolding in the minutes, days, and weeks following the initial injury. For starters, the brain, you'll recall, is a Jell-O–like bundle suspended inside a bone cage. If the outside of that cage—the skull—is met with force, the contents inside shift, often quickly and with force. So, the initial blow to Gary's head would have caused a "brain bruise"—a contusion at the point of contact called a coup—as it slammed into his skull. He might have also suffered a second brain bruise—called a contre-coup—in the part of his brain opposite the initial point of contact, as his brain rebounded from its initial hit and collided with the skull yet again. Often such injuries occur at the "tips" of the frontal and the temporal lobes, since these brain areas are closest to the skull. Severe TBIs lead to major brain bruising and swelling, which can cause further injury.

Many TBIs also cause "shear" or "diffuse axonal" injuries, in which the brain is subjected to a twisting motion inside the skull. As a result, the fiber bundles that carry signals across the brain are damaged. These are the brain's highways and, remember, they're not made of asphalt. In fact, they're more like a delicate collection of strands, each a thousand times thinner than the width of a human hair. It doesn't take an event as severe as Gary's baseball bat to the head to cause such an injury. Such damage can be done even when there is no direct contact at all, which might happen in a whiplash injury, for example. Shearing can happen with or without a brain bruise and can be either microscopic or large enough to be seen on MRI. Just as a silent stroke might go unnoticed, a small tear in a fiber bundle might cause no apparent symptoms, while a larger shear injury likely would produce noticeable issues. Depending on which part of the brain was damaged, cognitive problems that result might include not being able to calculate a tip in a restaurant, having difficulty finding your way in a familiar neighborhood, or struggling to complete a crossword puzzle, among many other possibilities.

Whether it's a contusion or a shear injury, if we zoomed in on neurons in the affected brain areas, we'd see injured mitochondria. These mitochondria would be unable to properly do their job, part of which is generating the energy needed for regulating the release of glutamate, an excitatory messenger molecule in the brain. With too much glutamate, affected neurons would be excited to death—a phenomenon called excitotoxicity. The more glutamate, the more damage.

Three Phases of TBI

INITIAL INJURY: Coup, contrecoup, and shearing.

SECONDARY INJURY: Inflammation and restricted blood flow.

REPAIR: Increased BDNF, synaptogenesis, and fiber sprouting.

The chain reaction caused by the TBI would continue as the brain's defense system leaped into action, sending specialized inflammatory cells to clean up damaged neurons. Unfortunately, while these cells do help, they also cause collateral damage when their cleaning efforts spill over to engulf healthy bystander neurons.

Collateral damage also includes injury to the blood vessel linings—called the blood-brain barrier—and allows chemicals to enter the brain.[1] This escalates inflammation, which then results in further damage to blood vessels and therefore reduced blood flow to the brain. As with glutamate, more inflammation means more damage. And it comes at the very time the brain needs optimal blood flow to fuel the healing process. The reduced blood flow—or ischemia—that results from such inflammation can continue to worsen over several days after the initial injury and can last for months.[2]

At the same time that the brain is fighting to limit damage, meanwhile, it is also beginning to repair itself. Immediately following an injury, production of the healing and building protein BDNF ramps up. Oftentimes, synaptogenesis increases and new dendrites sprout in the neurons of the injured brain.

Despite the brain's recovery efforts, though, the cascade often ends in brain atrophy, most obviously in the brain areas surrounding the injury but also in the hippocampus. In fact, seven studies conducted between 1997 and 2009 found a smaller hippocampus size in people who'd suffered either acute or chronic traumatic brain injury.[3] In my own review of the literature I was struck anew by the vulnerability of the hippocampus: even though the hippocampus doesn't take a direct hit, it is so sensitive that it shrinks, sometimes even more so than the areas that got bruised in the initial impact.[4]

Changes to the hippocampus can continue three to seven months after a TBI and are associated with cognitive decline. Even in children, who are blessed with a neuroplasticity that helps offset trauma, hippocampal shrinkage can persist later in life.[5]

Not surprisingly, given the havoc it wreaks, TBI significantly hampers healthy brain activity, which can persist long term if the injury is severe enough. In EEG brain maps of TBI patients I often see increased delta and theta activity in the frontal lobes—or "too much tuba." TBI patients also have low alpha and sometimes too much high beta activity as well. Depending on the level of injury, the "sounds" their orchestras produce may be anywhere from slightly off tempo to wildly chaotic.

Abnormal brain activity from TBI is so apparent, in fact, that a new, portable EEG device has been developed to help detect TBIs outside of a medical facility, be it on a football field, a battlefield, or anywhere else concussions are likely to occur. That device, called BrainScope, is now in clinical trial. Initial small trials showed that the handheld device can detect abnormal EEG patterns following a blow to the head, suggesting it may one day prove a valuable tool for coaches, parents, players, and soldiers.

Hard to Think

While it might seem obvious that injuries like Gary's would produce lasting cognitive problems, I often find patients are surprised to hear that brain injuries earlier in life may be contributing to their poor memories, confused thinking, or difficulty functioning in midlife. Even seemingly less severe, one-time events can cause headaches, balance issues, vision problems, ringing in the ears, and other physical problems that persist long past the injury. TBIs can also trigger mood disorders, such as depression, irritability, and anxiety, among other things.

Clearly major TBIs associated with a loss of consciousness shrink the brain—so much so that the effect can be seen on MRI. But even TBIs without noticeable symptoms can inflict microscopic damage that erodes brain reserve. Such "silent hits" may not result in a concussion diagnosis but may still affect brain function.

In one study, for example, a team of University of Illinois researchers had ninety college students perform cognitive tasks while undergoing EEG. Those who'd suffered concussions had slower electrical activity in their brains than those who hadn't suffered TBIs.[6] This, despite the fact that the injuries had occurred, on average, more than three years earlier and the students appeared outwardly to have no lasting effects from their injuries. In another study, students who'd had TBIs also showed slightly abnormal results on a test of balance, a finding that's often seen in the elderly at risk for falls.[7]

Other researchers have also put forth compelling evidence that multiple small injuries add up in the brain. While one big hit can create a major concussion, the damage from repeated minor traumas to the head can actually accumulate over time until a tipping point is reached, with the next small hit resulting in a sharp cognitive decline. In one study, for example, researchers at Purdue University looked at forty-five high school football players over two playing seasons.[8] Conducting cognitive testing and fMRIs, the researchers found that even silent hits caused changes in the brain, without causing noticeable symptoms. Brain reserve likely provided a cushion that allowed these young people to sustain damage to their neurons and highways and still function well. But brain reserve can eventually run low. It's not unlike the brake pads on your car: they're designed to cushion your brakes, but eventually, with enough mileage or strenuous use, they wear out.

At that point, damage can have far-reaching effects. As Gary discovered (and as researchers have demonstrated), injuries to the frontal lobes can cause damage that actually contributes to the likelihood of sustaining future head trauma. Even temporary symptoms can have an effect that might lead to another injury. A football player with a mild concussion, for example, might experience an almost imperceptible reduction in his reaction time while concussed, making him more likely to take a second hit.

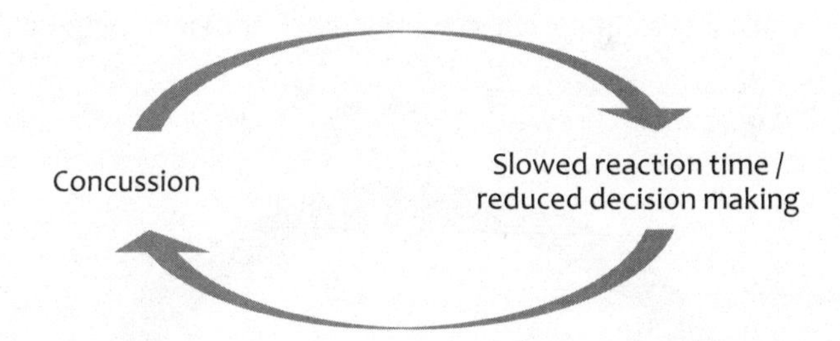

Even those who don't suffer another injury immediately are more likely to suffer future concussions. In one study, researchers followed 2,905 football players from twenty-five colleges and found that those who'd had a concussion were four to five times more likely to suffer another concussion in the future.[9] There's evidence, too, that recovery may be slower on subsequent concussions compared to the first.

Concussions also appear to increase the risk of dementia later in life. In 2010, a study of retired professional football players, for example, found that players who'd had three or more concussions were five times more likely to have cognitive impairment late in life than those without concussions.[10]

Another study found that losing consciousness for more than thirty minutes increases your risk of developing Alzheimer's disease later in life.[11] What's interesting about this is that a severe concussion—or repeated smaller hits—results in the immediate formation of the very tau tangles (and even amyloid plaques) that we see in the brains of patients with Alzheimer's disease.

Autopsy examination of young people who have died soon after TBIs shows a striking finding—despite their youth, their brains accumulated Alzheimer's pathology, both at the site of the brain contusion and in the hippocampus. This pathological finding, along with shear injuries to the fiber bundles and overall brain atrophy, explains why TBI patients can't think straight for years after their injuries.

One new brain imaging technique (FDDNP) that shows the presence of plaques and tangles in living patients with Alzheimer's disease has detected the very same lesions in young NFL players who've had concussions.[12]

The devastating formation of Alzheimer's pathology followed by significant impairments years or decades later is called chronic traumatic encephalopathy (CTE). This condition has been documented in boxers, military blast victims, and football players.[13] Sufferers of CTE experience brain damage that results in memory loss, confusion, depression, and aggression often years after the injuries are sustained.

Perhaps most alarming of all is evidence that the damage may begin to accumulate far sooner than was once thought and even in the absence of serious TBIs. It's a possibility not yet proven but certainly suggested by what

Brain Meter

Association with Brain Shrinkage

HIGH LOW

Multiple "silent hits," multiple concussions (with or without immediate symptoms), and more severe TBIs all are associated with brain atrophy. While the brain recovers from a few minor hits, the more frequent and the more severe the TBIs, the more your brain will shrink.

scientists found when they examined the brain of twenty-one-year-old college football player Owen Thomas, who committed suicide in 2010.[14]

Though Thomas had been a lineman, a punishing position that involves taking countless hits, he'd never had a documented concussion. The death of the once-happy, popular player was so inexplicable that his family agreed to donate his brain to the Center for the Study of Traumatic Encephalopathy at Boston University's School of Medicine in the hopes a closer look would offer answers.

What scientists found was shocking. Despite Thomas's young age and his lack of any serious head injuries, his brain showed the early signs of CTE.

Who Gets Hurt

Football gets a lot of attention for its head-damaging potential, but it's not the most common cause of TBI. The majority of TBIs, instead, come from car accidents or falls.

Still, many sports and recreational activities carry risks for TBI, at any age. Here are the top nineteen sports and recreational activities that contributed to head injuries in 2009, according to the American Association of Neurological Surgeons:[15]

Sport or activity	Emergency room visits per year
Bicycling	85,389
Football	46,948
Baseball and softball	38,394
Basketball	34,692
Water sports (diving, scuba diving, surfing, swimming, water polo, water skiing, water tubing)	28,716
Powered recreational vehicle use (ATVs, dune buggies, go-carts, minibikes, off-road vehicles)	26,606
Soccer	24,184
Skateboard and scooter use	23,114
Fitness and health club exercise	18,012
Winter sports (skiing, sledding, snowboarding, snowmobiling)	16,948
Horseback riding	14,466

Gymnastics, dance, and cheerleading	10,223
Golf	10,035
Hockey	8,145
Other ball sports, unspecified	6,883
Trampolining	5,919
Rugby and lacrosse	5,794
Ice skating	4,608
Roller and in-line skating	3,320

Getting Back in the Game?

There is no exact science behind how long it takes a bruised brain to heal completely. Part of the reason is that it's currently impossible to see microscopic damage that may have been done by a hit to the head.

For now, doctors rely on time and observation to tell them how well someone is healing. And until science comes up with a better way to measure how the brain is healing, it's best to tread with caution. Athletes with a suspected concussion, for example, should be taken out of play and shouldn't return until all symptoms have disappeared completely. Of course, this assumes that you have a diagnosed TBI. Oftentimes people don't actually get to that point because they brush off their symptoms or don't make the connection between their headaches and the hard knock they took, say, playing hockey the week before.

You can help minimize damage from a TBI by:

- seeking medical attention when a concussion or more serious TBI is suspected,
- following your doctor's orders for mental and physical rest following a TBI, and
- seeking out a specialist if problems persist.

Preventing Trauma

As someone who deals every day with people who've been laid low by traumatic brain injury, I have to admit that I've become pretty vocal about educating others on the importance of protecting the brain.

In many ways that's become easier, thanks to technologies like advanced safety devices in cars; seat belts and car seats for children; a more consistent use of helmets when biking, skating, skiing, or snowboarding; and a growing public understanding of the dangers of head injuries. In recent years coaches, parents, and doctors have begun to advocate for more protection for kids playing sports. New helmets offer better cushioning for the head, and safety-minded rules prevent some of the more dangerous moves in sports such as football. I'm hopeful that better awareness about concussion also means athletes of all ages will give themselves time to heal after a concussion before returning to play.

Many of my patients lead very active lives, and most want to continue to do so as they age. I'd like to do the same! Therefore, I don't tell them not to ski or cycle or even to stay out of the hockey rink. Instead, I tell them to give their brains a thought whenever they engage in any activity.

That might mean forgoing certain activities, such as boxing and mixed martial arts. In fact, I think boxing should be banned. It is, after all, traumatic brain injury in action! But giving your brain a thought might simply mean reducing your risk by wearing protective gear when playing sports or engaging in any activity where head trauma is a possibility. (Horseback riding; riding a motorcycle, snowmobile, or ATV; biking; skating; and skiing are a few.)

If you're thinking *nobody wears a helmet,* think again. Skiers and snowboarders aged eighteen to twenty-four—who traditionally have the lowest percentage of helmet use among all age groups—are increasingly donning helmets, according to the 2011 National Ski Areas Association's National Demographic Study. In 2011, 48 percent of all eighteen- to twenty-four-year-olds interviewed wore helmets, a 166 percent increase in usage for the age group since the 2002-2003 season, when only 18 percent wore head protection.

Given what I know about the rising risk of falls as we age, I also encourage regular safety checks around the home for anyone over the age of sixty-five. I ask them a series of simple questions. Are there loose rugs in your home, or electrical cords that might trip you? Could you benefit from a grab rail in the shower or elsewhere?

I also, of course, encourage all my patients to build their brain reserves, regardless of whether or not they've suffered a TBI (though, having had one is just another reason to devote effort to building up a nice, thick buffer).

Getting Better

Serious TBIs may be so obvious that diagnosis is a no-brainer. But diagnosing milder injuries can be more challenging. Symptoms can vary greatly by person: concussion sufferers may have migraines, memory loss, nausea, vomiting, insomnia, bowel problems, difficulty with concentration, inner ear problems that cause imbalance, sensitivity to light and sound, anxiety, irritability, or mood swings. Often such symptoms are dismissed outright or attributed to other causes. I frequently meet patients who've experienced cognitive symptoms for years without ever connecting them to a head injury they had suffered.

One, a young woman named Angela, spent years moving from doctor to doctor in search of answers to problems that cropped up after she suffered a concussion in a car accident. Once polished and professional, Angela found herself increasingly confused and unable to function at work following her concussion. She frequently felt dizzy and nauseated and began to suffer from daily migraines. Sometimes her pain and confusion were so severe that she would lock her office door and lie flat on the floor with her eyes closed. Before long, she added anxiety and depression to her list of woes. Within two years of her accident, she had lost her job and her marriage had crumbled.

When she came to see me nearly three years after her concussion, Angela was beginning to lose hope of ever regaining her health. To her, it was clear that the headaches and dizziness that followed her car accident were a direct result of her concussion. She suspected, too, that her mental confusion was related, but the handful of doctors she sought out for help dismissed her concerns. Some thought allergies might be her problem. Since her brain MRI was normal, her primary care doctor suggested she was simply an "anxious young woman."

She had, in fact, been through such a frustrating medical odyssey that when I told Angela her problems were very real and stemmed from her car accident, she put her head on my desk and cried with relief.

It's patients like Angela that make me recommend to anyone concerned about concussion that they find a doctor who is experienced in treating TBI patients.

That experience is crucial to treatment, too. Not surprisingly, given the variety of symptoms, there isn't a one-size-fits-all treatment option for concussion. Instead, each concussion symptom is addressed on its own. Migraine symptoms might be treated with medication, dietary changes, and treatment of insomnia. Anxiety or irritability might require counseling, stress-reduction training, or medication. Inner ear problems and gait imbalance might be corrected with physical therapy, and attentional difficulty might be treated with cognitive training. Often concussion patients are cognitively frail, so many of these treatments must proceed slowly and gently. But the end goal is the same: to treat symptoms and implement changes that boost BDNF, increase oxygen flow, and promote healthy brain activity. Put simply, these patients need to strengthen and grow their brains.

In the future, TBI patients may benefit from a number of treatment options now being researched. One possibility is progesterone, which may limit damage if given before an injury and speed recovery if given post-injury.[16] There's also work under way to determine if biomarkers in the bloodstream can be used to determine the severity of a TBI and gauge the effectiveness of treatment.

And there are signs that high doses of DHA given after an injury may also improve recovery.[17] In one study by my friend and colleague Dr. Julian Bailes, now chairman of the department of neurosurgery and co-director of the NorthShore Neurological Institute in Chicago, head-injured animals were treated with omega-3 fatty acids, including DHA, for thirty days following injury. Compared to a control group of injured but untreated animals, those given omega-3s had less damage in their brains. In fact, their brains looked much like animals who'd never had a head injury. There are now clinical trials in progress to examine whether DHA can do the same in humans and how much DHA is needed to produce the desired result.

Bailes and others are also working on another solution: a new helmet with a built-in accelerometer to detect the force of a hit. The helmet tracks how many hits a football player has taken and measures the force of the hits. Measurements are sent to a handheld device on the sidelines. Coaches and doctors can then track a player's accumulated hits—even silent hits—and assess the danger to the player.

Such tracking may be especially important given that in a majority of concussion cases CT scans or brain MRIs don't show obvious signs of injury, which can make assessing the damage difficult. EEG brain mapping may also be developed further to help us with establishing a firm diagnosis of concussion, especially when MRIs are normal.

These are all future solutions, however, so I encourage all my patients—those with TBIs and those without—to adopt a better-brain lifestyle that maximizes brain growers and minimizes brain shrinkers. For those who have experienced a TBI in the past, a healthy brain lifestyle is all the more important. It may be the difference between crossing the cognitive decline threshold and not.

Can it really help? For the answer, I'll take you back to Gary, whose cognitive skills were noticeably deficient for his age when he arrived for treatment and began his twelve-week brain fitness program. He was not physically fit at the start of his program—his fitness capacity, or VO_2 max, was just 59 percent—so getting him back in shape was a major goal. That job fell to my exercise physiologist, who worked with Gary to improve his VO_2 max by 5 percent every five weeks.

The physical effort helped to improve another of Gary's problems: insomnia, which is sometimes a consequence of TBI. Gary also worked with our clinical psychologist to further improve his sleep, and he learned to meditate and de-stress. Weekly EEG neurofeedback sessions, meanwhile, helped to increase his healthy brain activity and retrain his brain, as did computerized cognitive training sessions.

Although Gary's injuries had happened decades earlier, he still benefited from the marvel that is neuroplasticity. Three months after starting a brain fitness program, he had improved his cognitive scores significantly and reported feeling better than he had in years. Even his son, who'd been so discouraged at our first meeting, said his dad now seemed to have recovered to about 80 percent of his pre-illness level. It was a remarkable turnaround that spurred Gary's primary care physician to call me and ask, with pleasant surprise, "What did you do with my patient?" Gary was, as his son happily told me, "a new man."

CHAPTER FOURTEEN

Washing Away Your Brain's Neighborhoods and Highways

A FEW TIMES a month I see a new patient who comes in for "walking problems." One was Lara, a retired professor in her late sixties who arrived with her husband one afternoon for an appointment to discuss her difficulty with walking and balance. As I watched Lara approach, I could see she was unsteady on her feet: her gait was wide and wobbly.

As a part of my routine head-to-toe evaluation, I asked Lara about any cognitive issues she might have. It was then that her husband rolled his eyes and chimed in. "She doesn't remember anything!" he complained. "She's always repeating herself and asking me the same thing over and over."

Lara shrugged and she and her husband laughed good-naturedly about her lapses. It seemed, to them, a funny aside, but Lara's memory problems weren't really cause for a giggle. Nor were they, as she and her husband thought, completely separate from her unbalanced walking.

She didn't have vascular problems, wasn't overweight, and didn't have diabetes. It wasn't until I asked a key question that I had a good idea of what was at the root of Lara's problems. "Do you drink alcohol?" I asked.

She unabashedly answered yes, explaining that she regularly downed three to four stiff vodka cocktails a night. "How stiff?" I asked. "Really stiff," her husband answered. Lara, it turned out, had been guzzling vodka every night for thirty years.

She was rarely drunk and had never considered herself to have a drinking problem, but an MRI revealed that her drinking had indeed been a problem. I could clearly see that the part of her brain for balance and equilibrium, the cerebellum, was profoundly atrophied. Looking at it, Lara shook her head slowly. "But I'm always hearing news stories that alcohol is good for you," she said, genuinely dismayed.

Lara wasn't wrong. A little alcohol *is* good for most people (although it's not a significant brain booster and as such is low on my list of ingredients for a bigger brain). But crossing the line into alcohol abuse, perhaps even for short periods, quickly washes away those benefits. In fact, it shrinks the brain.

To the Top of the Curve, in a Bad Way

Lara could be forgiven for thinking that her nightly drinks might be beneficial. As you read in chapter 5, alcohol use, in moderation, has been shown to offer some neuroprotective benefit.

One well-regarded study published in the journal *The Lancet*, for example, studied a group of more than five thousand people over the age of fifty-five and found that those who drank one to three glasses of alcohol a day were significantly less likely—42 percent less—to develop dementia during the study period.[1] Just why isn't clear, although scientists have suggested alcohol in moderation may help to increase HDL cholesterol, which in itself is beneficial to the heart (and, thus, the brain).

Whatever the reason, alcohol's brain benefits disappear once you cross the line into alcohol abuse. Chronic, excessive alcohol use, for example, has been linked to a substantially greater risk of dementia later in life.[2]

Other studies have offered more insight into exactly what happens in the brain, as you'll read in a moment. Based on all the evidence, I consider alcohol beneficial in small doses and harmful in large doses. If you were to plot damage and alcohol use on a graph, for example, you'd end up with a J-shaped curve. Drink no alcohol and your brain will experience the usual degradation that happens with age; drink some and you'll see reduced damage; drink too much and injury to the brain shoots through the roof.

How Alcohol Harms the Brain

We know that alcohol in excess is bad for the brain. But why, exactly, is that? Does alcohol go straight to the brain and zap brain cells dead? Or is it, to go back to our CogniCity model, wiping out neighborhoods and highways alike? As it turns out, alcohol does its damage in several different ways.

To start, alcohol abuse does kill brain cells, and it affects certain parts of the brain more than others. One is the cerebellum, the part of the brain responsible for eye-hand coordination and balance, among other things. Unlike a cerebellar stroke, which causes immediate symptoms, alcohol damage in the cerebellum may result in walking problems that gradually worsen over time. Alcoholics may also have trouble with physical tasks that require coordination and precision, such as putting a car key into an ignition switch or moving a soup spoon from the bowl to the mouth.

Alcohol abuse also clearly shrinks the thinking and problem-solving parts of the brain. In one study that measured the brains of 130 alcohol-dependent people, for example, researchers found significant thinning of the cortex, compared to a control group.[3] In another study, researchers found alcoholics had less grey matter, especially in the prefrontal cortex.[4] That reduction correlated strongly with poorer scores on tests that measured decision making. In other words, the smaller their frontal lobes, the poorer their ability to make decisions. Alcoholics also had smaller hippocampi.

Sadly, poor decision-making abilities make it harder for alcohol abusers to make healthy choices, such as quitting alcohol, eating a balanced diet, exercising, or avoiding physical injuries. Alcoholics' gait problems may also increase their risk of falls, especially in old age, which may result in TBIs.

In addition, alcohol abuse damages myelin, the protective coating that covers neuron extensions. This causes neuropathy, or nerve damage, in the nerves leading from the spine to the feet. With their brains not getting good sensory signals from their feet, alcohol abusers suffer from poor proprioception—the body's sense of where its parts are in space. That combined with damage to the cerebellum leads to balance problems and difficulty walking. Lara's trouble walking was the early manifestation of this combination of factors. She did not yet experience falls, but in severe alcoholics such damage can cause frequent stumbling, particularly on stairs or curbs.

The brain of an alcohol abuser also suffers in another way: from the absence of key nutrients like thiamine, folate, and B12, caused by a poor diet. These nutrients are vital to overall brain function, and people who abuse alcohol often have dietary habits (beer and chips, anyone?) that deprive their brains of these important building blocks. Through poor diet, alcoholics may also short themselves on other brain-healthy essentials, like DHA.

Alcohol abuse wreaks its havoc elsewhere in the body, too, most notably in the liver, which can have a secondary effect on the brain. In long-term alcoholics, severe liver problems cause elevation of blood ammonia levels, which can lead to hepatic encephalopathy. This causes a variety of mental health issues, including shortened attention span, confusion, and tremors. Alcohol abuse is also tied to other brain-draining conditions, such as stroke and depression.[5] Eventually, alcoholics develop alcoholic dementia.

Some chronic alcoholics develop a serious disorder called Wernicke-Korsakoff syndrome, a disease that leads to vision problems, coordination problems, difficulty walking, disorientation, and learning and memory problems. These patients have severe brain atrophy, along with small hemorrhages in certain brain structures that are important for memory.

Your Brain on a Binge

If you think you're off the hook because you're not an alcoholic, but you sometimes overindulge in alcohol, you may want to reconsider. There's evidence that even binge drinking—heavy drinking that occurs sporadically—may do damage to the brain, especially if it occurs when the brain is still developing, in young adulthood. (Keep in mind that about 46 percent of young adults now report binge drinking,[6] so the practice is strikingly common.)

In one study of 122 Spanish college students aged eighteen to twenty, for example, researchers at the Universidade de Santiago de Compostela found that binge drinkers differed from their non-bingeing peers when it came to performance on cognitive tests.[7] As compared to sixty non-binge-drinking men and women, sixty-two binge drinkers had difficulty with performance in tests of attention, fluency, and abstract design. They also showed evidence of perseveration, which reduced mental flexibility. Researchers concluded that these results may indicate frontal lobe dysfunction or developmental delay in the binge drinkers.

These findings were bolstered by an imaging study that showed actual damage on MRIs. This study, presented to the Research Society on Alcoholism in mid-2011 by University of Cincinnati researchers, looked at the brains of twenty-nine eighteen- to twenty-five-year-olds who engaged in weekend binge drinking.[8] MRIs of the binge drinkers' brains showed a thinning of the prefrontal cortex.

Fortunately the study also offered evidence that abstaining from alcohol after binge drinking allowed the brain to recover, an outcome we can chalk up to the beauty of plasticity. But since we don't have good studies showing the long-term effects of binge drinking—or how much the brain recovers after binge drinking—I recommend erring on the side of caution.

Have a Smoke?

You know smoking damages the brain. But we're beginning to understand more about just where in the brain smoking hits hardest. We now know that smoking thins part of the frontal lobe called the orbitofrontal cortex, an area that's been tied to both impulse control and rewards.[9] Heavier smokers have pronounced thinning in this area, suggesting smoking may take a cumulative toll on the brain. This might also explain, in part, why smokers find it so hard to quit: they've damaged the very part of the brain they need to help them control their impulse to light up.

What does that mean, exactly? I tell all my healthy patients that one or two glasses of alcohol (one for women, two for men) per day is probably beneficial, as long as they don't suffer from any memory or thinking problems. But anything more is risky, and regular heavy drinking is riskiest of all. For people with memory complaints, I advise avoiding alcohol altogether. Even a drop of alcohol is too much, I tell them.

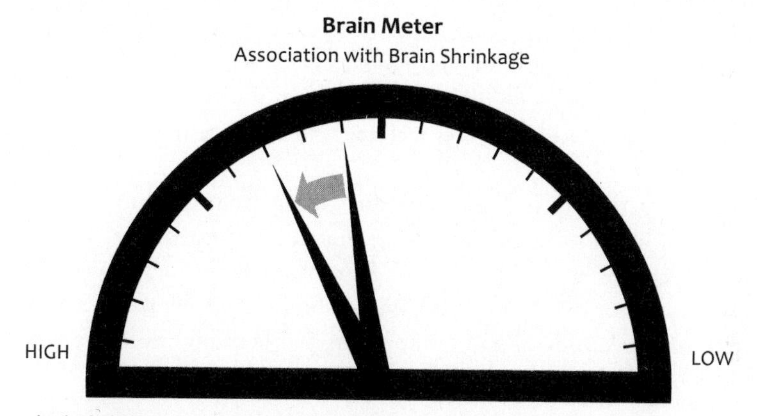

Alcohol abuse is associated with brain shrinkage. A long-term habit of consuming more than four servings a day of alcohol increases the damage; the more you consume, the worse the damage.

Getting Better

When Lara left my office she was committed to ditching the alcohol, and reviving her brain. Although moderate alcohol use has benefits, I advised her (as I advise all my patients with memory problems) to gradually reduce her alcohol intake and then avoid alcohol entirely. She joined my brain fitness program and with the help of a psychologist, an exercise physiologist, and an EEG neurofeedback trainer, over the next twelve weeks she lowered her alcohol consumption, improved her fitness, stimulated her memory, and retrained her brain.

Fortunately for Lara and others like her, some of the damage done by alcohol can be rapidly undone. One study, for example, followed fifty patients admitted for an alcohol withdrawal program in Germany and found that three months of abstinence seemed to reverse some of the damage to the brain.[10] When first evaluated, study subjects' MRIs showed brain shrinkage. Three months later (during which time patients abstained from alcohol), MRIs showed parts of the brain, especially the cingulate gyrus—important for emotion and mood—had grown in size. Patients who relapsed during the study period didn't see such volume growth.

Another study conducted by some members of the same research team found that just two weeks of abstinence produced measurable—but not complete—reversals of volume loss in several brain areas.[11] This is yet another example of the brain's remarkable neuroplasticity and capacity to quickly heal itself.

In fact, abstinence brings about a burst of regeneration of brain cells, as one study in animals showed.[12] In that study, animals were given ethanol to mimic four days of binge drinking, which, as expected, resulted in atrophy in the hippocampus. After abstinence, though, researchers were able to see the birth of new neurons across the hippocampus. This was a striking difference compared to the animals who continued to receive alcohol.

In my experience, alcoholic patients who stop drinking and work to improve brain health often see their walking and memory improve within six to eight months.

Fortunately for my patient, Lara, abstaining from alcohol and engaging in extensive brain training resulted in dramatic improvement. In her favor, she had the advantage of not having significant other factors—obesity, high blood pressure, or diabetes—shrinking her brain. And her years of cognitive stimulation as a professor had no doubt helped her enter her later years with a healthy brain reserve.

Got a Problem?

Evaluate your drinking habits honestly. Do you have a drinking problem? If a sincere assessment of your drinking leads you to wonder whether it falls in the unhealthy range, the National Institute on Alcohol Abuse and Alcoholism suggests asking yourself these four questions:

- Have you ever felt you should cut down on your drinking?
- Have people annoyed you by criticizing your drinking?
- Have you ever felt bad or guilty about your drinking?
- Have you ever had a drink first thing in the morning to steady your nerves or to get rid of a hangover?

If you answer yes to one or more of these questions, you may be abusing alcohol. A health care professional should be able to help you seek further assessment or help.

It's especially important to be under the care of a doctor as you stop drinking. Stopping abruptly can trigger seizures, so it should be done only under the close supervision of a medical professional. After a gradual period of detoxification, recovering alcoholics typically need to spend time in rehabilitation and then will need support to maintain sobriety.

PART V

On the Horizon

CHAPTER FIFTEEN
Is It Alzheimer's?

I'M RARELY ALONE at parties. I often find myself surrounded by a circle of family or friends who pepper me with questions about memory and aging. Sometimes they're merely curious about the latest research in the field, but often it's more personal than that: they're worried about what's going on in their own brains.

Inevitably someone will launch into a discussion of his latest lapse, sheepishly describing the time he walked out of a mall, bogged down with shopping bags, and spent thirty minutes wandering the rows of sedans and SUVs in an increasingly desperate search for his car. Or the sinking feeling he had when he realized—too late—that he'd missed an important meeting. Or how he will purposefully stride into a room and then forget, the moment he passes through the doorway, what he came for. Eventually, he'll nervously joke, "I swear, I'm getting Alzheimer's. Right?"

It's a common fear. In fact, in a Shriver Report poll conducted in 2010 and reported in *TIME* magazine, 84 percent of respondents said they were concerned that they or someone in their families would be affected by Alzheimer's disease.

But is it a reasonable fear? For the vast majority of the people who ask me if they have Alzheimer's disease, the answer is no. They don't have the disease but instead are experiencing the normal mild memory problems and slowed mental processing that can come with age.[1] A small number of people, however, may well be at the start of cognitive decline that has at its root a mix of factors and may eventually lead to late-life dementia. And an even smaller number—a tiny fraction—of the people I meet in social situations are experiencing the start of a pure form of Alzheimer's disease that strikes people in their fifties and sixties. How do we know who has what? And is there anything we can do to alter the course? The truth is that answering these questions is a little more complicated than you might think (but it's fascinating!).

What Party?

Remember Sara, the baby whose brain development we followed in chapter 1? Imagine Sara as a sixty-year-old. At this stage of life it would not be unusual for her to experience age-associated memory impairment, a relatively benign condition that occurs naturally as we age and results in minor memory lapses and a slight reduction in the brain's processing speed. That slowing would explain why Sara's grandchildren might handily beat her at the game of Whac-a-Mole, which requires quick reaction times, but not at chess, which relies more heavily on strategy, planning, and experience. Sara might misplace her keys or experience tip-of-the-tongue syndrome, where a word she's seeking seems maddeningly just out of reach. But with age-associated memory impairment, she wouldn't experience a serious decline in her thinking or her daily functioning.

If Sara's thinking or memory problems exceeded the level deemed normal for her age, she might be diagnosed with *mild cognitive impairment* (MCI), a middle ground between normal brain aging and dementia. With MCI, Sara might regularly forget conversations she had with her husband (and be met with "I told you that four times already!") but she would still function fairly well at work and at home. Sara could still drive, dress herself, cook, clean, and pay bills. She would, however, have an increased risk of developing dementia.

Dementia is a broad umbrella term used for people with serious brain atrophy causing profound and progressive decline in memory, along with problems in one or more other areas of cognition such as decision making, calculations, or orientation. If Sara developed dementia, she might start out with mild memory loss but quickly progress to confusion and, eventually, an inability to function. Dementia can have many causes, including stroke, life-long alcohol abuse, viral infections, severe TBI (as in boxers), HIV, or neurodegenerative conditions such as Alzheimer's disease, Parkinson's disease, frontotemporal dementia, or Lewy body dementia, in which different sets of brain cells die for unknown reasons. If Sara's doctors couldn't pinpoint a single cause but determined that more than one factor contributed to her dementia, she might be diagnosed as having "mixed dementia," although such a diagnosis isn't used anywhere near as much as I believe it should be (as you'll soon read).

If Sara's doctors didn't find specific evidence of the cause of her dementia, she would likely be diagnosed with *Alzheimer's disease,* a subset or type of dementia (although people often use the terms dementia and Alzheimer's interchangeably). With Alzheimer's, as with many other forms of dementia, Sara might experience profound memory and cognitive problems as well as personality and behavioral changes. She might get lost in her own neighborhood, forget how to tie a shoelace, or be unable to operate a simple device like a can opener. She might be sad for no reason, or cry or fly into a rage without provocation. Though she might remain physically healthy, Alzheimer's is a progressive, degenerative disease, so it inevitably worsens and leads to death. If her brain were examined after her death, it would likely be full of amyloid plaques and tau tangles, the hallmark toxic protein clumps of Alzheimer's disease.

To illustrate the difference between memory deficits seen in normal aging, mild cognitive impairment, and Alzheimer's disease, we can think about what might happen a few days after one of those parties I attend.

If I were to bump into a person with benign age-related memory problems and ask "who did you talk to at the party last weekend?" I might get a response along the lines of "I spent a lot of time with that woman I always see in the lunch room at work. She's tall, blonde, and very friendly. I know she told me, but I can't remember her name."

A person with mild cognitive impairment, on the other hand, might say, "I talked to so many people. I just have no idea who!"

A person with Alzheimer's, though, would have a strikingly different answer. When asked about the party they'd attended just a few days before, this person might look genuinely perplexed and pose a question of her own: "What party?"

What Causes Alzheimer's

As telling as the "what party?" response is, it doesn't actually identify for us specifically what's going on inside the brain of the sufferer. "What party?" tells us there's a problem but not why it occurred.

So, why *does* Alzheimer's occur? The answer depends on what you mean by Alzheimer's. There are actually two distinct types of Alzheimer's: early-onset and late-onset.

In early-onset Alzheimer's disease, which often strikes people in their fifties and sixties, we know exactly what causes the brain shrinkage and decline associated with the condition. Alzheimer's pathology—amyloid plaques and tau tangles—wreak havoc, rapidly killing brain cells. And while plaques and tangles don't show up on commonly used brain scans, brain MRIs of early-onset Alzheimer's patients do show marked atrophy in the hippocampus in the early stages, followed by atrophy scattered throughout the brain. I think of it as a forest fire. The match may be dropped in the hippocampus, but the flames quickly spread—with devastating results.

You might consider early-onset Alzheimer's a "pure" form of the disease because it has solely Alzheimer's pathology, rather than the consequences of lifestyle factors or medical conditions, at its root. This form of the disease is largely determined by genetic makeup and it accounts for a small subset of total Alzheimer's cases. How small? In the United States an estimated two hundred thousand people have early-onset Alzheimer's.[2] Given that about five million people have Alzheimer's disease, we can assume the portion of those having early-onset Alzheimer's is about 4 percent.

The other 96 percent have late-life Alzheimer's disease. The brains of these people exhibit the same shrinkage in the hippocampus and across the brain, but they also often harbor clues pointing to other causes. In fact, in the brains of people who develop Alzheimer's disease in late life we often see signs of both silent strokes and atrophy caused by other neurodegenerative conditions, such as Lewy body disease, as well as any number of health problems—obstructive sleep apnea, hypertension, diabetes, and depression, to name a few.[3]

In contrast to early-onset Alzheimer's disease, in late-life Alzheimer's disease genetics make up a small part of the risk. To put that genetic contribution into perspective, consider this: while the average seventy-year-old man has a roughly 2 percent chance of developing late-life Alzheimer's disease, someone of the same age who has close relatives with late-life Alzheimer's might have a 4 percent chance. That's double the chance, sure, but just like buying two lottery tickets instead of one doubles your likelihood of winning (but still leaves you highly unlikely to walk away with millions), doubling your risk of Alzheimer's disease at this level is hardly a death knell.

A far larger portion of the risk of developing late-life Alzheimer's disease is tied to a group of factors we have some degree of control over. Give that seventy-year-old multiple vascular risk factors, such as high blood pressure, a BMI greater than 30, physical inactivity, and high total cholesterol, and his risk of developing late-life Alzheimer's disease rockets to 32 percent.[4]

Many studies have examined the link between clinical symptoms of dementia and the primary pathology in the brain.[5] In one—with the telling title "Mixed Brain Pathologies Account for Most Dementia Cases in Community-Dwelling Older Persons"—researchers at Chicago's Rush University Medical Center carefully examined the brains of 141 deceased elderly people and came to the conclusion that many did not have just a single abnormality.[6] The most common abnormality the research team noted was a mix of plaques, tangles, small strokes, large strokes, and Lewy bodies (Parkinson's-like lesions). In fact, those with dementia had a load of plaques and tangles similar to those without dementia; what was different was the presence of other pathological components in what I call a "soup of factors." Thus, the best diagnosis for such patients would be "mixed dementia," although in most cases they are instead labeled as having Alzheimer's disease.

In chapter 16, I'll address the controversy over terminology surrounding late-life Alzheimer's disease. But for now it's enough to know that when it comes to dementia that strikes in late life I consider Alzheimer's pathology yet another brain-shrinking factor, one among the soup of other factors, such as diabetes, obesity, stroke, TBI, alcohol abuse, stress, poor sleep, and depression.

The Tricky Business of Diagnosis: Silent Alzheimer's

Though we have amassed an impressive array of technologies related to the brain, we still don't have a simple, reliable, and definitive method of determining if someone has Alzheimer's disease. Unlike diabetes, which we can detect through a blood test, or hypertension, which we identify through blood pressure readings, there is no single test for Alzheimer's disease.

Often diagnosis, then, is tricky. This is less true in the case of "pure" early-onset Alzheimer's disease, which tends to be fairly straightforward. Patients who develop dementia in their fifties and sixties often experience a rapid decline in their memories and ability to function without signs of the soup of other factors—a tip-off that Alzheimer's might be the sole cause.

By contrast, late-onset Alzheimer's patients tend to see their cognitive decline progress more slowly and their neurological evaluations often turn up hints of other conditions—like vascular disease—that contribute to shrinkage in the brain. Making diagnosis more difficult, this soup of brain shrinkers varies in every person. Some may have a few ingredients; others may have many. The evidence may be clear (like a major stroke) or almost invisible (like thousands of micro strokes).

Until fairly recently, the only way to confirm Alzheimer's disease was by examining the brain after death. If the person suffered from dementia while alive and had an accumulation of plaques and tangles in the brain, that person would have a confirmed case of Alzheimer's disease. But even then there was, and still is, room for variation. Some doctors make their diagnoses based solely on the prevalence of amyloid plaques; some consider amyloid plaques and tau tangles; still others give tau tangles the lion's share of the weight. There is no single "gold standard" in diagnosing Alzheimer's (even after death), which, as you can imagine, leads to great discrepancies in who is diagnosed and with what.[7]

Even Alzheimer's experts in academic centers may have different opinions when it comes to diagnosis. I experienced this firsthand in the early 2000s when I was a consultant for the Johns Hopkins Alzheimer's Disease Research Center, which convened weekly meetings with a group of neurologists, psychiatrists, and pathologists. Together we would discuss patients who, after being monitored for twenty or thirty years, had recently died. Hearing about their symptoms and test results, we'd each make an educated guess as to their diagnoses, based on the pathological findings. Was it mild cognitive impairment? Alzheimer's? Lewy body dementia? Vascular dementia? In many cases, individual members of the team disagreed on the diagnoses. Although that was a decade ago, the lack of consensus among experts persists, despite the aid of improved technologies.

Such discrepancies may be, in part, because some elderly people have "silent" Alzheimer's in their brains. Just as silent strokes can cause damage without leading to symptoms, "silent" Alzheimer's pathology—those plaques and tangles—can also do harm without leading to noticeable symptoms. It's only when combined with other factors that a threshold is reached and cognitive decline begins.

This might explain why the Alzheimer's *pathology* we've long associated with Alzheimer's disease doesn't always correlate well with late-life Alzheimer's *symptoms*. Some people have Alzheimer's plaques and tangles and never develop symptoms, while others have the same load of pathology and become demented. Why? The reason likely has to do with the coexistence of silent strokes, concussions, or medical conditions as well as the strength of the person's brain reserve. Each brain shrinker chips away at the brain's buffer; when multiple shrinkers are combined, the deficits overcome the brain reserve.

One often-cited piece of evidence for this is the so-called Nun Study, which began in the 1980s and followed the health and lifestyle of 678 American members of the School Sisters of Notre Dame.[8]

At its heart, the study aimed to answer the question of why some elderly people suffer from dementia while others do not. What was their secret? How did the über-sharp ninetysomethings among us do it? The nuns, it was hoped, would offer an answer.

As part of the study, participants agreed to have their brains examined after they died. When considered along with detailed cognitive studies and MRIs taken while the nuns were alive, the autopsied brains offered up some surprising clues. Nuns whose brains had Alzheimer's tau tangles but no sign of strokes typically hadn't displayed any outward symptoms of Alzheimer's disease while they were alive. Similarly, many of those who'd had multiple strokes—large or small—but few tangles in their brains did not show symptoms of Alzheimer's disease. But those who had tangles *and* evidence of strokes had a strong likelihood of displaying Alzheimer's symptoms. Alzheimer's pathology, it seemed, wasn't the sole cause of late-life Alzheimer's disease for most nuns in the study.

Given that symptoms likely reflected a mix of factors, which do you blame? Should we point the finger at tangles? Or silent strokes? Or were there other factors too?

Does it really matter what we call it? Actually, it does, as the loved ones of any recently diagnosed Alzheimer's sufferer would tell you. Alzheimer's disease is often seen as heartbreakingly unalterable—a death sentence.

I often favor using the term "mixed dementia" in cases where I believe multiple etiologies have contributed to a patient's decline. This would help to ensure that only those whose Alzheimer's symptoms are caused primarily by Alzheimer's pathology would be diagnosed as having Alzheimer's disease.

Such a distinction has a benefit beyond just the impact on patients. In research studies, focusing just on patients who have significant plaques and tangles—and excluding those whose dementia had other causes—would allow scientists to be more targeted than ever. An amyloid-busting drug would offer little benefit to someone whose dementia was 80 percent stroke-related and only 20 percent caused by plaques and tangles, for example. Removing

Brain Meter

Association with Brain Shrinkage

HIGH LOW

The plaques and tangles of Alzheimer's Disease, especially in young patients, can rapidly destroy the brain.

**With aging, most elderly people accumulate some degree of plaques and tangles in their brain without having any symptoms; this is called "silent Alzheimer's." Some will accumulate a great deal more and experience serious brain atrophy and dementia. The more plaques and tangles, the more their brains shrink and the more likely they are to develop memory loss, confusion, or difficulty in managing their simple daily routine.*

†When combined with other brain shrinkers—like diabetes, hypertension, obesity, sedentary lifestyle, depression, PTSD, TBI, and sleep apnea—the risk of brain atrophy escalates and the patient experiences dementia faster.

that person from the Alzheimer's disease category would allow researchers to instead study just those who might benefit from amyloid removal. In fact, this may be the key reason that all the amyloid-busting drugs have failed so far: they've been tested on too broad a group.

A diagnosis of mixed dementia, when properly applied, could open up new treatment options for patients who might otherwise be seen as untreatable. That's especially true given that reducing cardiovascular risk factors and preventing future strokes are known to be important strategies for preventing or slowing the progression of mixed dementia, as one group of researchers recently noted in *JAMA*.[9]

Diagnosis In Action

Although there's no single test to determine if someone suffers from pure Alzheimer's disease, new procedures do exist to identify the presence of Alzheimer's pathology in living people. For now, such methods aren't routinely employed by most doctors for diagnosis and are not recommended by the American Academy of Neurology.[10] Doctors, instead, rely on a battery of tests—some cutting edge, some tried and true and decidedly low-tech. Those methods might include MRI scans, blood tests, and a variety of health measures to rule out other conditions, as well as cognitive testing to gauge a person's level of impairment.

Here are some of the diagnostic tests now available:

CSF Cerebrospinal fluid (CSF) tests can look for a specific ratio of tau to amyloid, which seems to be highly indicative of a high risk for Alzheimer's disease. This test isn't perfect, as some people with abnormal CSF may not have symptoms of dementia. In addition, it's somewhat invasive, as it requires taking a sample of spinal fluid through lumbar puncture.

PET Scans These are new brain imaging techniques that can detect, by using radioactively labeled ligands (or markers), the presence of specific proteins in the brain. Radiologists have developed new ligands to label and visualize the footprints of Alzheimer's disease in a living patient. These ligands are:

- *FDG*, a marker for glucose. Low FDG in the temporal, parietal, and frontal lobes is indicative of low brain activity and may indicate damage caused by Alzheimer's pathology.
- *Pittsburg compound B* (PiB), which allows us to see amyloid plaques in the brain. Unfortunately, this ligand only alerts us to the presence of amyloid; it doesn't tell us about tangles and it doesn't tell us who will develop dementia and how severe it will be. In fact, between 20 and 50

percent of elderly people might have a positive PiB test without showing clinical signs of dementia.[11]

- *Florbetapir,* a newer ligand that detects the presence of amyloid. It is now available to the general public, but Florbetapir suffers from the same disadvantage as PiB. Some 21 to 28 percent of cognitively healthy elderly people tested have Florbetapir results that indicate the presence of amyloid even though the person doesn't have dementia.[12] In addition, a third of the patients who have symptoms of Alzheimer's disease have negative Florbetapir results.
- *FDDNP* (fluorine fluoroethyl [methyl] amino naphthyl ethylidene malononitrile), which detects both plaques and tangles.[13] FDDNP may prove to be the best tool for diagnosing Alzheimer's, but the downside is that its results aren't specific to Alzheimer's. It can also pick up other conditions that are known to cause plaques and tangles, such as in football players who have chronic traumatic encephalopathy (see chapter 13), leading to the possibility that a football player in his sixties (who had concussions in his twenties) might be misdiagnosed with Alzheimer's disease.[14] Boxers and soldiers with blast trauma would also likely have a positive result on this test.

Of course, a negative scan using any of these methods can be extremely helpful because it tells us amyloid and possibly tau *aren't* present.

Brain MRI Of all the tests currently available, I favor brain MRIs that measure hippocampal size and atrophy. These tests are noninvasive and the results are quite robust, thanks to the strong correlation between Alzheimer's disease and shrinkage of the hippocampus. Still, they're not perfect. Some people with TBI or depression may have small hippocampi. If their symptoms are caused by these factors, we wouldn't want to diagnose them as having Alzheimer's disease.

Based on the fact that no one test is 100 percent accurate, a group of experts has suggested that an Alzheimer's diagnosis should be given only if a patient has dementia plus abnormal results on one or more of the tests above (CSF, PET, or MRI for hippocampal atrophy).[15] This is a work in progress and I expect new developments in the near future.

Slowing the Slide?

If you were experiencing memory or other cognitive problems late in life and you knew or suspected you had Alzheimer's disease, you and your loved ones would likely be eager to find a cure. Or at least a way to slow your decline. And your children would likely be anxious to see what they could do to prevent themselves from developing the disease in old age. Since Alzheimer's pathology clearly plays a role, one option would be to seek out a cure that clears the brain of plaques and tangles or prevents their further accumulation. Unfortunately, though there's been much research in this area, neuroscientists haven't had a great deal of luck so far. Part of the reason is that plaques and tangles likely accumulate for decades before dementia appears, making late-life treatment "too little, too late." So far, every clinical human trial aimed at removing amyloid plaques has failed to produce a cognitive benefit for patients. There is some hope, however. New trials show some positive results that we may build on in the future.

Another option would be to realize that Alzheimer's pathology isn't the only cause of late-life Alzheimer's disease in most cases. This means you might be able to grudgingly accept its presence and focus instead on eliminating other brain-shrinking ingredients in the soup of dementia causes.

The scientific evidence for prevention is quite strong. Hundreds of research studies, some of which you've already read about, suggest modifying certain factors reduces the risk of developing Alzheimer's disease late in life.

Those efforts almost always come down to boosting brain size. We know, for example, that having a large hippocampus greatly reduces your likelihood of experiencing the symptoms of dementia, even if you have the footprints of Alzheimer's in your brain.[16] Therefore, actions that boost hippocampal size are vital to remaining free of symptoms. Increasing BDNF, enhancing oxygen flow, and promoting healthy brain activity—as you'll do in your twelve-week brain fitness program—will help add synapses and blood vessels, and bolster highways throughout the brain.

But could such interventions help people who already have MCI or Alzheimer's disease? Although little can be done to reverse the damage of advanced Alzheimer's disease, recent studies suggest that certain interventions may provide some help for MCI patients.

In one randomized controlled clinical trial in Seattle, for example, researchers enrolled thirty-three adults with MCI and assigned half to a stretching group and half to a high-intensity aerobics group.[17] The aerobics group exercised under the supervision of a fitness trainer for forty-five to sixty minutes a day, four days a week, for six months. The control group, meanwhile, completed stretching activities on the same schedule. All participants underwent treadmill fitness testing and memory testing before and after the study. Six months after they'd started, those who'd engaged in aerobic exercise showed improvements on tests of executive function and had increased their levels of BDNF.

Resistance training may also help. In one study, researchers found that resistance training promoted cognitive and functional brain plasticity in patients with MCI.[18] The study included eighty-six women between the ages of seventy and eighty. By the close of the six-month study period, those who'd engaged in resistance training showed improved performance on tests of executive function and functional changes in the frontal lobes, as captured by fMRI.

Cognitive stimulation has also been shown to bring with it subtle brain performance improvements in MCI patients. One randomized controlled pilot trial of forty-seven MCI patients, conducted by my colleague Dr. Kristine Yaffe, looked at the effects of intensive computer-based cognitive training.[19] The training was specifically designed to improve auditory processing speed and accuracy and was given to patients for a hundred minutes a day, five days a week, over a period of six weeks. A control group, meanwhile, engaged in passive computer activities, such as reading or listening. Participants in the cognitive training group, whose average age was seventy-four, saw small improvements in their total testing scores. And, compared to the control group, they did slightly better on verbal learning and memory after training.

Another small randomized study of forty-three people with MCI and mild Alzheimer's disease found that a group assigned to cognitive stimulation for six months saw improvements in memory and mood.[20]

A large systematic review of this literature, published in 2012, concluded that MCI patients do benefit from cognitive stimulation—and that positive changes can be detected on MRI.[21] This review of twenty studies from different research centers around the world provides evidence that neuroplasticity is at work even in patients who have MCI.

Prevention and Treatment

You've no doubt heard talk about an epidemic of Alzheimer's disease on the horizon. In 2007, the Johns Hopkins Bloomberg School of Public Health announced its prediction that the number of people suffering from Alzheimer's disease worldwide would quadruple by 2050. Even now, the Alzheimer's Association puts our risk of developing Alzheimer's disease by the age of eighty-five at nearly 50 percent.[22] Those are pretty grim statistics. But what do they mean for you? Can you prevent Alzheimer's disease? The answer, once again, depends on what we mean by Alzheimer's disease.

For early-onset Alzheimer's disease the answer is no, at least for now. Researchers may one day discover a way to silence the gene mutation responsible for the disease or block the aggregation of plaques and tangles, but currently there isn't much that can be done in the way of prevention. Three large clinical trials with anti-amyloid drugs are about to start, however, so we may learn more as they wrap up, beginning in 2016.

The news is much brighter for late-life Alzheimer's disease. Researchers will continue their efforts to find a cure (my prediction is that there will never be a single cure), but in the meantime, preventive measures come down to building brain reserve and reducing the factors that shrink the brain. This was the basis of my first book, *The Memory Cure* (published in 2002), and though the notion was controversial at the time, now it is widely accepted.

This all, of course, comes with a caveat: even the most rigorous of efforts can't prevent every case of late-life dementia. There will always be people who eat right, exercise, sleep well, and treat their health conditions and yet still go on to develop Alzheimer's disease late in life. But for most people prevention is entirely possible. And as someone who sees the effects of Alzheimer's disease every day, I can tell you it's a worthwhile endeavor.

At my Brain Center, my goal is to help patients sharpen their brain performance and prevent any memory loss or further cognitive decline in the future. But the goal posts move depending on the patient's situation, and the lines between treatment and prevention blur, since some treatments can act as prevention, helping to slow further decline.

For each patient in my brain fitness program, I develop an individualized treatment plan that is specific to them, their needs, and their families' needs. As you've read throughout the book, the effects can be dramatic, even if the program is started in middle age or late life.

Back to the Party

Remember those questions from worried friends and family at social gatherings I attend? My immediate response, "It's probably not Alzheimer's," always earns me a relieved smile. But as happy as my friends and family are to hear they're not likely suffering from a dreaded disease, they're not off the hook. In fact, the memory lapses and slowed thinking that prompted their concern are a powerful reminder of the importance of a brain-healthy lifestyle throughout life. Not only will these efforts yield nearly immediate results, but the brain reserve created in the process will also go a long way toward preventing, or at least delaying, development of the type of late-life dementia that's often diagnosed as Alzheimer's disease.

CHAPTER SIXTEEN

Back to the Future

YOU'VE READ this far—and sprouted some new synapses as you soaked up chapter after chapter of the latest science—so you know the powerful impact of brain boosters and shrinkers that shape your brain throughout life. You know that multiple vascular risk factors can lead to strokes so small they cannot be measured on MRI or so large they cause instant, catastrophic damage. You know that multiple concussions can create silent "tears" that produce subtle changes in your brain function; that years of downing stiff drinks can quietly kill neurons and erode the brain's highways; that sleep apnea can rob the brain of oxygen, taking a toll on grey and white matter alike. You know that all of these factors—plus a few more—can reduce your memory, clarity, and creativity long before you reach old age. And you know that any of them can combine with the presence of plaques and tangles in the brain to tip you toward dementia late in life.

In the future, we will no doubt have at our disposal ever more amazing tools to help us harness brain growers and avoid brain shrinkers. We will be better equipped than ever to measure and track brain size and health, modifying our behaviors as needed. We'll do it all with the end goal of enhancing our brain performance now and preserving it in the future, knowing that cognitive decline in late life is anything but inevitable.

Such a mind-set seems obvious now, but that hasn't always been the case. In fact, the way we view brain size and health, especially in late life, has changed dramatically over the years. So, before I tell you about the exciting developments on the horizon, let's take a journey through the centuries to see just how we arrived at our current outlook on brain aging.

A Walk Through History

Long before we developed the tools to peek inside the human brain, it was recognized that old age brings with it a decline in cognition. We didn't need imaging technologies or complex cognitive tests to figure this out. Greek philosopher and mathematician Pythagoras noted in the seventh century BC that the aged return to "the imbecility" of infancy. Later, Hippocrates called mental decline an inevitable consequence of aging—a fate he blamed on an "imbalance of fluids." Aristotle, too, weighed in, noting that the elderly were gradually blunted by mental deterioration and suggesting that the heart was the center of sensation and movement (and that the brain's job was to cool the heart).[1]

It wasn't until midway through the second century AD that the physician and philosopher Galen suggested the brain was the home of cognition and the center of mental disorders, including dementia. Science largely took a backseat to religion during the Middle Ages, a time when dementia and other disorders were seen as punishment for sins—payback for a lack of piety.

As the seventeenth century dawned, however, scientific heads prevailed. With religious taboos ebbing, dissection became an acceptable way to explore the mysteries of the human body. Thomas Willis, who had developed an atlas of the blood vessels in the brain, coined the term "neurology" and drafted a classification of dementia, which included in its very modern list of causes aging, head injury, and alcohol and drug abuse.

By the 1800s, scientists had documented atrophy in the brains of deceased demented people and had begun to delve deeper into its potential causes. Many felt syphilis was to blame. Others were not so sure. Researchers at the time were aided by microscopes, which had been in wide use for scientific discovery since the 1600s. But it wasn't until 1893 that August Köhler unveiled a new technique to illuminate samples, ushering in a new era of modern light microscopy. A rush of discovery would soon follow, including the first description of gummy black clumps (plaques) in the brain of an elderly epilepsy patient in 1898 and the first identified case of dementia caused by white matter disease.

When a well-respected young German psychiatrist and neuropathologist named Aloysius "Alois" Alzheimer took on a curious case in the early 1900s, the stage for discovery was set. The patient, known to us only as "Auguste D.," was a German woman in her early fifties who had rather suddenly begun acting strangely. She was confused and disoriented and had difficulty speaking and remembering her past. Confined to a mental institution, Auguste intrigued Dr. Alzheimer, who had an intense interest in both forensic pathology and dementia. In an elderly patient, such symptoms might be ascribed to "senile dementia"—the term of the day for late-life dementia—but in someone so young? What could it be?

Dr. Alzheimer meticulously documented Auguste's rapid decline over the next five years. When she died, in 1906, he autopsied her brain.

At his disposal he had a newly enhanced mode of microscope and a just-created silver staining technique—two wonders that when combined offered, for the first time, a look inside the cellular structure of the brain. Peering at slices of Auguste's brain, Dr. Alzheimer noted the presence of plaques. But he also saw something new: tangled threadlike structures within the neurons in Auguste's brain. Some cells had just a few tangles, while others seemed to contain bundles of them. Still other neurons had disintegrated completely, replaced by masses of tangles strewn about Auguste's brain. All told, as many as a third of the neurons in Auguste's brain had been obliterated.

By this point, Dr. Alzheimer and the rest of the scientific community had a good feel for how the brains of the very old looked. But Auguste was different. She was, after all, not yet sixty.

Dr. Alzheimer published a brief report, calling the case "a peculiar disease of the cerebral cortex." Over the next few years he sought out other patients who'd died after having exhibited similar symptoms. Examining their brains he found the same evidence of plaques and tangles.

It wasn't just their ages that struck Dr. Alzheimer. The brains of these young demented patients looked distinctly different from the brains of their older counterparts. In the young, plaques and tangles mottled the brain; in the elderly, they were present but were accompanied by other signs of trouble, chiefly atherosclerosis, or hardening of the arteries, which shrunk the brain in a process then known as "senile involution." Other scientists had already demonstrated that the brains of the demented shrink. But Dr. Alzheimer had been one of the first to describe stroke, or "the gradual strangulation of the blood supply to the brain," as the cause of such atrophy. As he described the factors at work in the demented elderly, Dr. Alzheimer was careful to note his belief that plaques weren't the cause of senile dementia "but only an accompanying feature of senile involution of the central nervous system," as he wrote in 1911.[2] To Dr. Alzheimer the evidence made it clear: dementia in young patients was quite different from dementia in the elderly.

The discovery didn't create much of a stir at the time. Alzheimer was, after all, just a fortysomething researcher with a handful of unusual cases. The disease might never have gotten his name if it weren't for the chairman of his department, Emil Kraepelin, a renowned psychiatrist and author. Kraepelin had played a part in describing and defining the diseases we know now as schizophrenia and bipolar disorder and by 1911 was in the midst of writing the eighth edition of the diagnostic manual of the day, *Textbook of Psychiatry.* Hoping to one-up his peers at a competing university in Prague, Kraepelin decided to include Dr. Alzheimer's cases, giving the condition its now-famous name.

What Dr. Alzheimer would have done in the years to come is anybody's guess. He died unexpectedly in 1915 at the age of fifty-one. And while he'd gained prominence and respect for his life's work, none of those who eulogized him thought fit to describe him at the time as the person who discovered Alzheimer's disease.

A Pendulum Swings . . .

It would be many decades before anyone really took note of Alzheimer's disease. The supposition by Dr. Alzheimer and others that vascular disease caused senile dementia held well into the century that followed, as did use of the term "hardening of the arteries" to describe dementia in the elderly.

In fact, vascular disease was so central to the view of late-life dementia that papaverine, a vasodilator that opens up blood vessels and was widely prescribed for dementia patients, became one of the most oft-prescribed medications in the United States.

By the middle of the century, multi-infarct dementia—dementia whose primary cause was multiple strokes—had been identified and the importance of inflammation in the development of brain pathology had been suggested. To be sure, the presence of plaques and tangles in the brains of senile dementia patients, and even in those who'd died without any symptoms of dementia, sparked some debate. But for the most part, all eyes were on vascular disease.

As health care improved, vascular problems that would have once killed people outright in their sixties and seventies were suddenly better understood and controlled (although not, it should be noted, completely), ushering more people into their eighties and nineties—and dementia. By the 1950s and 1960s, diagnosed cases of senile dementia were on the rise.

At the same time, technological advances allowed us a closer look into the workings of the brain. With the public clamoring for an answer to the vexing problem of dementia in the elderly, science took a renewed interest in the findings of Alois Alzheimer.

Homing in on the brains of those with late-life dementia, scientists documented marked shrinkage in the hippocampus and broad stretches of cerebral cortex. They found, too, that in the brains of those with late-life dementia the neurotransmitter important for memory (acetylcholine) was in short supply and that plaques in the brain were made of a protein called beta amyloid. Searching for genetic clues, they discovered the specific gene mutations in the amyloid protein that caused the formation of toxic amyloid plaques in early-onset Alzheimer's disease. Soon after, scientists discovered the protein at the core of tangles, called *tau*.

What followed was a sea change in the perception of dementia. These were, after all, the same amyloid plaques and tau tangles found in patients with early-onset Alzheimer's disease. Since demented patients late in life also had plaques and tangles, they should all fall under the broad umbrella of Alzheimer's, unless there was some obvious other cause. Or so the thinking went.

The unwavering focus on Alzheimer's pathology—those plaques and tangles in the brain—from the 1980s on seemed to swing the pendulum of scientific thought away from a focus on vascular disease and its role in late-life dementia.

At the time, from 1988 to 1992, I was a graduate student at Johns Hopkins University working with Dr. Solomon Snyder, one of the most prominent neuroscientists of our time. Next door was one of the pioneers in Alzheimer's research, Dr. Donald Price, who was then director of the Johns Hopkins Alzheimer's Disease Research Center.

I vividly recall the excitement in the hallways as Dr. Price's army of graduate students and lab technicians zeroed in on the genetics and physiology of amyloid plaques. In the early 2000s, the thrill of the race to cure Alzheimer's disease ratcheted up yet again as interest focused on the "amyloid cascade hypothesis," which suggested that amyloid was the chief cause of Alzheimer's disease. The idea had been proposed a decade earlier, but suddenly it was hot.[3]

Under this theory, the damage seen in Alzheimer's disease—shriveled synapses, dead neurons, shrinkage in the hippocampus, loss of overall brain volume, and disappearance of key neurotransmitters like acetylcholine—all had amyloid plaques at their root. Even neurofibrillary tangles were believed to be downstream damage caused by amyloid plaques, secondary to the inflammation that occurred as the body attempted to defend itself from excess amyloid.

Not everyone was sold on the idea. Still, the hypothesis generated immense enthusiasm in the medical community. If Alzheimer's disease was caused by amyloid plaques then curing it might simply be a matter of developing a drug to prevent amyloid plaques from forming or to clear the brain of amyloid plaques once they formed. Right? A decade later—and after several failed trials—we know it's not nearly so simple. Amyloid plaques, as Dr. Alzheimer said in 1911, are only part of the picture.

. . . And Swings Back

As you read this book, the pendulum is swinging once again toward a consensus that vascular disease is one of the chief culprits in late-life dementia. One reason for the shift is that we now have overwhelming evidence of the link between vascular disease and shrinkage in the brain with aging. The second reason is that we've accumulated evidence that the older we get, the weaker the link is between Alzheimer's pathology and symptoms of dementia. In fact, by the time we're in our eighties and nineties, the presence of plaques and tangles isn't strongly tied to the development of dementia.[4]

In one study published in the *New England Journal of Medicine,* for example, researchers in the United Kingdom examined 456 donated brains from people aged sixty-nine to a hundred and three and found that the older the people were when they died, the weaker the relationship between amyloid plaques and dementia was.[5] It's the opposite of what you'd expect to see if amyloid plaques alone were at the heart of Alzheimer's disease: as they aged, demented study subjects had *lower* levels of amyloid plaques in their brains than those who weren't demented.

The third reason for the shift away from an emphasis on plaques and tangles as the sole cause of dementia in the elderly is the discovery that older people may accumulate a significant load of Parkinson's-type pathology (called Lewy body lesions) in their brains in addition to plaques, tangles, and vascular damage. Dr. Lon White, the principal researcher of the long-running Honolulu-Asia Aging Study, has intensely explored the link between clinical dementia and pathological findings in brain autopsies. In an interview with *Neurology Today,* White noted that the majority of the elderly patients he studied have a combination of vascular lesions, Lewy body lesions, and Alzheimer's lesions.[6] In fact, only 22 percent of brain autopsies showed plaques and tangles as the sole cause of dementia. It's an important reminder, Dr. White said, that "there is an enormous overlap" in the causes of late-life dementia. David Knopman, a Mayo Clinic professor of neurology, took it one step further, telling *Neurology Today* that "our idea of thinking about Alzheimer's disease as a single clinicopathological entity simply doesn't work." Alzheimer's disease, in other words, is not just a simple accumulation of plaques and tangles. There are other culprits at work.

For anyone worried about late-life dementia, that's astoundingly good news. Vascular disease is, by and large, preventable. If we can reduce our risk for developing it, we also reduce our risk of being felled by the late-life condition we call Alzheimer's disease.

A "Dynamic Polygon"

I have a great deal of respect for Dr. Alzheimer. He, after all, knew the disease he noted in Auguste D. differed from that which struck the very old. To reflect that reality, and the fact that the words "dementia" and "Alzheimer's disease" are so thoroughly stigmatized, I—along with two prominent neuroscientists, Dr. Vladimir Hachinski and Dr. Peter Whitehouse—favor a radical change in terminology.[7]

Together, we have suggested replacing the terms dementia and Alzheimer's disease with the more respectful and less stigmatizing terms: mild, moderate, or severe "cognitive impairment." We've also put forward an alternative to the amyloid cascade hypothesis, with an emphasis on the fact that most cases of cognitive impairment in the elderly are caused by numerous different etiologies and that these factors exist in a dynamic state in the brain, each capable of ramping up injury caused by the others. According to our "dynamic polygon hypothesis," sleep apnea, insomnia, diabetes, obesity, depression, and stroke act both individually and in combination with Alzheimer's plaques and tangles (or Lewy body lesions) to chip away at the brain. Those brain-shrinking culprits are balanced by brain reserve created by multiple brain-boosting factors, such as exercise, meditation, cognitive stimulation, a heart-healthy diet, and high levels of DHA.

Our hypothesis suggests that given the dynamic interaction of various factors involved in causing brain atrophy, treating one condition (such as hypertension) may reduce levels of another factor (such as plaques and tangles in the brain). A new discovery published in 2013 by Dr. White and his team supports our hypothesis. It shows that a subset of blood pressure medications (called beta blockers) that improve blood flow to the brain reduce not only the number of strokes in the brain but also the load of Alzheimer's plaques and tangles.[8]

This concept is further supported by the results of research by Dr. John Morris and his colleagues at Washington University in St. Louis.[9] Using an amyloid PET imaging technique, they first confirmed the high density of amyloid in the brains of patients with a genetic disposition that increases the risk for Alzheimer's disease (called APOE ϵ4). They then found that patients with this inherited mutation who exercised regularly had a low brain amyloid level, comparable to unaffected people who didn't have this genetic mutation. In other words, it appears these individuals were able to use exercise to reverse the effects of an Alzheimer-related APOE ϵ4 mutation they were born with.

These exciting breakthroughs, yet to be replicated, point to the fact that we may be able to directly prevent late-life Alzheimer's disease through the same interventions that prevent strokes and heart attacks. It's yet more incentive to engage in brain-boosting activities throughout life, especially if you have family members who have Alzheimer's disease.

Looking to the Future of Brain Health

I consider myself lucky to be in a field that is in the midst of unprecedented discovery. We are, in fact, entering a new era in brain health, one that will bring with it a dramatic shift in the way we view the brain throughout life.

With the explosion of interest in the field of neuroscience and the urgent need to stop age-related cognitive decline and Alzheimer's disease, we can expect to see a surge of new revelations and technologies that address the factors affecting brain size with aging.

One area of particular interest to me is the effect of our increasingly wired lifestyle on brain size. We already know, as I discussed in chapter 7, that

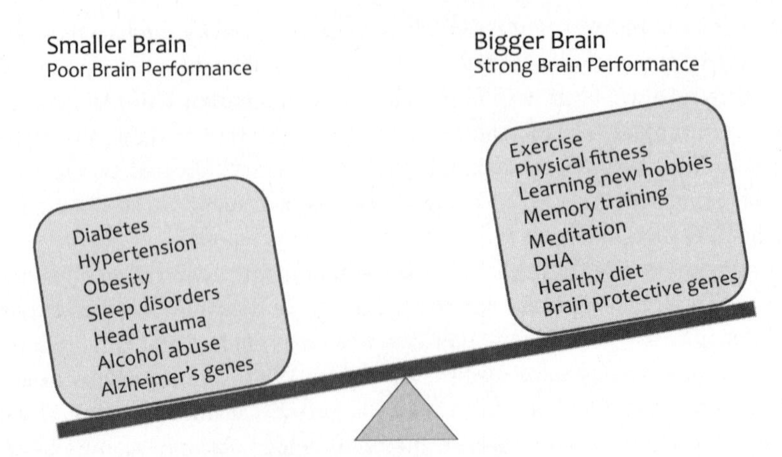

Your brain health and performance are determined by a balance of factors that either shrink or grow the brain. Genes play a small role in late-life brain health: just as you may have genes that increase your risk of Alzheimer's disease, you may also have protective genes that keep you mentally sharp.

paying close attention and using our memory and other cognitive skills are important for maintaining those skill sets. I expect we'll discover more about how frantically checking e-mail, being glued to electronic devices, and being in the habit of skimming information reshapes the brain. As we rely less on our brains to deeply digest the information that comes our way, to calculate math problems, or to memorize phone numbers, will the generations to come have smaller hippocampi? And to what effect, on both brain function and their future risk of developing late-life dementia? I don't know the answer, but I do fear our patterns of technology use, constant distraction, and over-stimulation will come at a cost.

On the other hand, technology will no doubt play a beneficial role in many ways. We may soon see, for example, devices that measure and monitor our "brain health index" (based on EEG recordings), our "brain fitness index" (based on our capacity to handle challenging cognitive tasks or on test performance), our "brain hit index" (based on sensors inside helmets), or our "brain size index" (based on quantitative MRI measurements of the hippocampus). Already I routinely obtain detailed information about hippocampal volume in my patients because I know the direct link between its size and the potential for remaining sharp for years to come. We might one day also measure the quantity of synapses in the brain ("brain synapse index") and use that measure as an indicator of brain health and fitness throughout life. Just as you know how much money you have in your retirement account, you may one day know how many synapses you have in your "brain reserve account."

I believe that the advance in imaging techniques will revolutionize our views of aging and what we define as "normal" versus "Alzheimer's disease." With new PET labeling markers (such as FDDNP) that show the presence of plaques and tangles in living people,[10] we will need to decide if these pathological clumps of toxic amyloid plaques are the culprits themselves or just the consequence of things gone wrong. Are they the fire or merely the smoke? It's a vitally important distinction, since removing the smoke won't stop a fire. If plaques and tangles are merely a reflection of damage to the brain, then perhaps the term "Alzheimer's disease" will take on a new meaning. This has tremendous implications for diagnosing and labeling patients who have memory issues.

As exciting as the future of technology and testing is, there's plenty of action ahead in the world of food and drugs as well. We're already aware of the importance of eating well, but we'll continue to learn more about how certain foods—from blueberries to broccoli, clams to coconut oil—affect the brain and how much we need to consume to reap rewards.

Emerging brain "superfoods" may one day contain a cocktail of herbs and supplements, such as DHA, curcumin, quercetin, resveratrol, vitamin B12, folate, Fruitflow, and huperzine. There is also a wealth of research under way into various pharmaceutical treatments for late-life dementia. Countless drug trials have failed to find a cure, but some hold promise.

I'm not terribly hopeful that a single cure—pharmaceutical or otherwise—will be found for Alzheimer's disease and other late-life dementias. But I am hopeful about a future in which we have far better options to control various known brain drainers throughout our lives. There's much excitement around potential "cures" for obesity, for example, through interventions that help to regulate the hormone leptin, which is critical for controlling appetite.

With a greater understanding of how to stave off brain atrophy, it's likely that just as we have experienced an increase in life span over the past century, we will see an increase in our "brain span"—the portion of our lives that we live in peak cognitive condition. The bigger your brain size, the greater will be your brain span.

The Face Behind the Facts Is . . . You

Remember the story from chapter 4 of my co-author, Christina? Not long after we began working together on this book, she announced she wanted to put my brain fitness program to her own (unscientific) test.

Like any other patient in my program, she completed neurocognitive testing, took blood tests for key health measures, and took a fitness test to gauge her cardiovascular health, followed by an MRI to check for any brain abnormalities and to measure the size of her hippocampus. Not too surprisingly given her youth and health history, her test results showed she was in good health, with no indicators of any problems, save a slight vitamin D deficiency.

For the next three months, she dove headfirst into her brain fitness efforts, rekindling a long-stalled running routine, dropping her worst eating habits, practicing relaxation exercises, and adding DHA to her diet. She made efforts to reduce stress and kept her brain active by immersing herself in brain-related research. As we talked over her brain fitness scores each week, it was clear that she was doing well. She shed ten pounds, began to sleep better, and reported an increase in her energy level. She even completed a half marathon.

But what about her brain? As you'll recall from chapter 4, Christina's hippocampus grew by a whopping 5 percent. This is the equivalent of sparing her ten years of brain aging. Not only that, but when she underwent formal cognitive re-evaluation at the three-month mark, her performance had improved in two separate tests of memory, jumping 16 percent in her speed of problem solving and 15 percent in memory performance. In short, she remembered more and performed faster.

It was a breathtaking result and something I see again and again at my Brain Center. And while Christina's is just one story, it stands as my reminder that behind all the dry scientific studies, behind the statistics, behind the carefully controlled trials, there are real people. Given the right tools, as I've seen countless times in my medical practice, they can counter the worst effects of brain aging and improve their memories, their clarity, and their creativity within months.

And what about you? By now you've either implemented or are about to get started on your own twelve-week plan to boost your brain. I'm confident you'll have noticeably enhanced brain performance to show for it. What then? Though you'll be wrapping up your three-month program, don't view it as the end. Instead, consider the habits you've developed as a new beginning. Make growing your brain a part of your every day, working fitness, a healthy diet, mindfulness, quality sleep, and cognitive stimulation into everything you do. And make reducing brain shrinkers your mission. Carry those habits forward and you'll reap the rewards of a bigger, stronger brain for years to come.

Imagine that. Will it happen? It can. You now know that you have the power to change your brain. What happens next is up to you.

Acknowledgments

EARLY ON in the process of writing my first book, *The Memory Cure,* I knew that it wouldn't be long before I'd once again tackle the topic of memory and brain health. I also knew that when I did, luck would be on my side if I had two key players in my corner: HarperOne editor Nancy Hancock and literary agent Anna Ghosh. Their guidance and support, along with the help of HarperOne's Elsa Dixon and Dianna Stirpe, were invaluable in bringing *Boost Your Brain* to life.

The staff of my Brain Center's brain fitness program—on which this book is based—along with my executive team, David Abramson, Steve Dubin, Jeff Hamet, Chris Lindsay, and Petuna Selby, all gamely pitched in on book-related duties, offering advice, critiques, and support at every turn. I am particularly thankful to Dr. Eylem Sahin for reading the full manuscript carefully and providing us with detailed comments. My star pre-med student Brooke Lubinski helped at every stage of the game and on multiple occasions rushed to our aid to unearth an obscure reference, whip up a chart, or just brainstorm—invaluable assistance that made the task immeasurably easier, and this book better. Baltimore-based artist Ashley Milburn rounded out our corps of helpers, patiently sketching and re-sketching until neurons practically danced off the page.

I'd also like to thank a host of neuroscience and other experts, who spent time detailing their research and sharing their knowledge. They include: Paula Bickford, Michelle Carlson, Vincenza Castronovo, Kevin Crutchfield, Richard Davidson, Charles DeCarli, Helene Emsellem, Kirk Erickson, Fred Gage, Vladimir Hachinski, Sara Lazar, Bruce McEwen, Norman Salem, and Molly Wagster, along with Joseph Maroon and Gertjan DeKoning at DSM. I thank my dear friend the two-time U.S.A. Memory Champion, Nelson Dellis, for sharing his memory tricks with us.

Above all, I am most thankful to my coauthor Christina Breda Antoniades. She worked tirelessly and with great enthusiasm in crafting a compelling book proposal, thoroughly researching each chapter, and interviewing my colleagues to get firsthand accounts of their research. Through our hundreds of conversations, she patiently listened and made sense of my explanations of a vast array of complex neurology concepts—from the molecular mechanisms involved in brain development in the uterus to the latest brain imaging technologies. Ever the journalist, she volunteered to put herself through our twelve-week brain fitness program and experienced all the tests and procedures at my Brain Center (along with the benefit of a bigger, stronger brain) in order to best understand how and why the program works. I thank my friend, DSM's Cassie France-Kelly, for introducing Christina to me.

Finally, this project would never have made it past the planning stage if it weren't for the backing of our families. Heartfelt thanks to my wife, Bita, and to Christina's husband, Spiro, whose patience and understanding allowed us to complete this book.

Notes

Introduction

1. Majid Fotuhi, Vladimir Hachinski, and Peter J. Whitehouse, "Changing Perspectives Regarding Late-Life Dementia," *Nature Reviews Neurology* 5, no. 12 (2009): 649–58.

2. Majid Fotuhi, David Do, and Clifford Jack, "Modifiable Factors That Alter the Size of the Hippocampus with Aging," *Nature Reviews Neurology* 8, no. 4 (2012): 189–202.

3. D. Erten-Lyons, R. L. Woltjer, H. Dodge, R. Nixon, R. Vorobik, J. F. Calvert, M. Leahy, T. Montine, and J. Kaye, "Factors Associated with Resistance to Dementia Despite High Alzheimer Disease Pathology," *Neurology* 72, no. 4 (2009): 354–60.

Chapter 1: Your Marvelous Mind

1. www.youtube.com/watch?v=mMDPP-Wy3sI.

2. Nitin Gogtay, Jay N. Giedd, Leslie Lusk, Kiralee M. Hayashi, Deanna Greenstein, A. Catherine Vaituzis, Tom F. Nugent III, David H. Herman, Liv S. Clasen, Arthur W. Toga, Judith L. Rapoport, and Paul M. Thompson, "Dynamic Mapping of Human Cortical Development During Childhood Through Early Adulthood," *Proceedings of the National Academy of Sciences of the United States of America* 101, no. 21 (2004): 8174–79.

3. Elizabeth R. Sowell, Paul M. Thompson, and Arthur W. Toga, "Mapping Changes in the Human Cortex Throughout the Span of Life," *Neuroscientist* 10, no. 4 (2004): 372–92.

Chapter 2: How to Grow a Brain

1. Majid Fotuhi, Vladimir Hachinski, and Peter J. Whitehouse, "Changing Perspectives Regarding Late-Life Dementia," *Nature Clinical Practice Neurology* 5, no. 12 (2009): 649–58.

2. Majid Fotuhi, David Do, and Clifford Jack, "Modifiable Factors That Alter the Size of the Hippocampus with Aging," *Nature Reviews Neurology* 8, no. 4 (2012): 189–202.

3. http://cogns.northwestern.edu/cbmg/plasticityInGMandWMreview2012.pdf; www.jneurosci.org/content/28/35/8655; www.ncbi.nlm.nih.gov/pubmed/21906988.

4. Emma G. Duerden and Danièle Laverdure-Dupont, "Practice Makes Cortex," *Journal of Neuroscience* 28, no. 35 (2008): 8655-57.

5. Gerd Kempermann, H. George Kuhn, and Fred Gage, "More Hippocampal Neurons in Adult Mice Living in an Enriched Environment," *Nature* 386, no. 6624 (1997): 493–95.

6. Peter Eriksson, Ekaterina Perfilieva, Thomas Björk-Eriksson, Ann-Marie Alborn, Claes Nordborg, Daniel Peterson, and Fred Gage, "Neurogenesis in the Adult Human Hippocampus," *Nature Medicine* 4, no. 11 (1998): 1313–17.

7. Heidi Johansen-Berg, Cassandra Sampaio Baptista, and Adam G. Thomas, "Human Structural Plasticity at Record Speed," *Neuron* 73, no. 6 (2012): 1058–60.

8. D. Erten-Lyons, R. L. Woltjer, H. Dodge, R. Nixon, R. Vorobik, J. F. Calvert, M. Leahy, T. Montine, and J. Kaye, "Factors Associated with Resistance to Dementia Despite High Alzheimer Disease Pathology," *Neurology* 72, no. 4 (2009): 354–60.

9. Ian J. Deary, Jian Yang, Gail Davies, Sarah E. Harris, Albert Tenesa, David Liewald, Michelle Luciano, Lorna M. Lopez, Alan J. Gow, Janie Corley, Paul Redmond, Helen C. Fox, Suzanne J. Rowe, Paul Haggarty, Geraldine McNeill, Michael E. Goddard, David J. Porteous, Lawrence J. Whalley, John M. Starr, and Peter M. Visscher, "Genetic Contributions to Stability and Change in Intelligence from Childhood to Old Age," *Nature* 482, no. 7384 (2012): 212–15.

Chapter 3: Creating Your Twelve-Week Plan

1. www.heart.org/HEARTORG/Conditions/Cholesterol/AboutCholesterol/What-Your-Cholesterol-Levels-Mean_UCM_305562_Article.jsp.

2. Heidi Johansen-Berg, Cassandra Sampaio Baptista, and Adam G. Thomas, "Human Structural Plasticity at Record Speed," *Neuron* 73, no. 6 (2012): 1058–60.

3. Josef Shargorodsky, Sharon G. Curhan, Gary C. Curhan, and Roland Eavey, "Change in Prevalence of Hearing Loss in U.S. Adolescents," *Journal of the American Medical Association* 304, no. 7 (2010): 772–78.

4. www.aoa.org/x5253.xml.

5. N. Okamoto, M. Morikawa, K. Okamoto, N. Habu, J. Iwamoto, K. Tomioka, K. Saeki, M. Yanagi, N. Amano, and N. Kurumatani, "Relationship of Tooth Loss to Mild Memory Impairment and Cognitive Impairment: Findings from the Fujiwara-Kyo Study," *Behavioral and Brain Functions* 31, no. 6 (2010): 77.

6. M. A. Woo, P. M. Macey, G. C. Fonarow, M. A. Hamilton, and R. M. Harper, "Regional Brain Gray Matter Loss in Heart Failure," *Journal of Applied Physiology* 95, no. 2 (2003): 677–84.

7. O. P. Almeida, G. J. Garrido, C. Beer, N. T. Lautenschlager, L. Arnolda, N. P. Lenzo, A. Campbell, and L. Flicker, "Coronary Heart Disease Is Associated with Regional Grey Matter Volume Loss: Implications for Cognitive Function and Behaviour," *Internal Medicine Journal* 38, no. 7 (2008): 599–606.

8. Rea Rodriguez-Raecke, Andreas Niemeier, Kristin Ihle, Wolfgang Ruether, and Arne May, "Brain Gray Matter Decrease in Chronic Pain Is the Consequence and Not the Cause of Pain," *Journal of Neuroscience* 29, no. 44 (2009): 13746–50.

9. J. L. Phillips, L. A. Batten, F. Aldosary, P. Tremblay, and P. Blier, "Brain-Volume Increase with Sustained Remission in Patients with Treatment-Resistant Unipolar Depression," *Journal of Clinical Psychiatry* 73, no. 5 (2012): 625–31.

Chapter 4: The Fit-Brain Workout

1. Henriette van Praag, Gerd Kempermann, and Fred Gage, "Running Increases Cell Proliferation and Neurogenesis in the Adult Mouse Dentate Gyrus," *Nature Neuroscience* 2, no. 3 (1999): 266–70.

2. Kirk I. Erickson, Ruchika S. Prakash, Michelle W. Voss, Laura Chaddock, Liang Hu, Katherine S. Morris, Siobhan M. White, Thomas R. Wójcicki, Edward McAuley, and Arthur F. Kramer, "Aerobic Fitness Is Associated with Hippocampal Volume in Elderly Humans," *Hippocampus* 19, no. 10 (2009): 1030–39.

3. Kirk I. Erickson, Michelle W. Voss, Ruchika Shaurya Prakash, Chandramallika Basak, Amanda Szabo, Laura Chaddock, Jennifer S. Kim, Susie Heo, Heloisa Alves, Siobhan M. White, Thomas R. Wojcicki, Emily Mailey, Victoria J. Vieira, Stephen A. Martin, Brandt D. Pence, Jeffrey A. Woods, Edward McAuley, and Arthur F. Kramer, "Exercise Training Increases Size of Hippocampus and Improves Memory," *Proceedings of the National Academy of Sciences of the United States of America* 108, no. 7 (2011): 3017–22.

4. Laura Chaddock, Kirk I. Erickson, Ruchika Shaurya Prakash, Jennifer S. Kim, Michelle W. Voss, Matt VanPatter, Matthew B. Pontifex, Lauren B. Raine, Alex Konkel, Charles H. Hillman, Neal J. Cohen, and Arthur F. Kramer, "A Neuroimaging Investigation of the Association Between Aerobic Fitness, Hippocampal Volume, and Memory Performance in Preadolescent Children," *Brain Research* 1358 (2010): 172–83.

5. Éadaoin W. Griffin, Sinéad Mulally, Carole Foley, Stuart A. Warmington, Shane M. O'Mara, and Áine M. Kelly, "Aerobic Exercise Improves Hippocampal Function and Increases BDNF in the Serum of Young Adult Males," *Physiology and Behavior* 104, no. 5 (2011): 934–41.

6. K. I. Erickson, C. A. Raji, O. L. Lopez, J. T. Becker, C. Rosano, A. B. Newman, H. M. Gach, P. M. Thompson, A. J. Ho, and L. H. Kuller, "Physical Activity Predicts Gray Matter Volume in Late Adulthood: The Cardiovascular Health Study," *Neurology* 75, no. 16 (2010): 1415–22.

7. S. J. Colcombe, K. I. Erickson, P. E. Scalf, J. S. Kim, R. Prakash, E. McAuley, S. Elavsky, D. X. Marquez, L. Hu, and A. F. Kramer, "Aerobic Exercise Training Increases Brain Volume in Aging Humans," *Journals of Gerontology, Series A: Biological Sciences and Medicine Sciences* 61, no. 11 (2006): 1166–70.

8. C. Raji, K. Ericson, O. Lopez, J. Becker, O. Carmichael, H. M. Gach, P. Thompson, W. Longstreth, and L. Kuller, "Energy Expenditure Is Associated with Gray Matter Structure in Normal Cognition, Mild Cognitive Impairment, and Alzheimer's Dementia," paper presented to the Radiological Society of North America on November 26, 2012.

9. E. Bullitt, F. N. Rahman, J. K. Smith, E. Kim, D. Zeng, L. M. Katz, and B. L. Marks, "The Effect of Exercise on the Cerebral Vasculature of Healthy Aged Subjects as Visualized by MR Angiography," *American Journal of Neuroradiology* 30, no. 10 (2009): 1857–63.

10. Stefan Schneider, Christopher D. Askew, Thomas Abel, Andreas Mierau, and Heiko K. Strüder, "Brain and Exercise: A First Approach Using Electrotomography," *Medicine and Science in Sports Exercise* 42, no. 3 (2010): 600–7.

11. "Montreal Study: Sport Makes Middle-Aged People Smarter," Université de Montréal Nouvelles, October 29, 2012, www.nouvelles.umontreal.ca/udem-news/news/20121029-montreal-study-sport-makes-middle-aged-people-smarter.html.

12. Frédéric N. Daussin, Joffrey Zoll, Elodie Ponsot, Stéphane P. Dufour, Stéphane Doutreleau, Evelyne Lonsdorfer, Renée Ventura-Clapier, Bertrand Mettauer, François Piquard, Bernard Geny, and Ruddy Richard, "Training at High Exercise Intensity Promotes Qualitative Adaptations of Mitochondrial Function in Human Skeletal Muscle," *Journal of Applied Physiology* 104, no. 5 (2008): 1436–41.

13. J. Z. Willey, Y. P. Moon, M. C. Paik, M. Yoshita, C. DeCarli, R. L. Sacco, M. S. Elkind, and C. B. Wright, "Lower Prevalence of Silent Brain Infarcts in the Physically Active: The Northern Manhattan Study," *Neurology* 76, no. 24 (2011): 2112–18.

Chapter 5: Your Recipe for a Bigger Brain

1. M. L. Corrêa Leite, A. Nicolosi, S. Cristina, W. A. Hauser, and G. Nappi, "Nutrition and Cognitive Deficit in the Elderly: A Population Study," *European Journal of Clinical Nutrition* 55, no. 12 (2001): 1053–58.

2. Nikolaos Scarmeas, Yaakov Stern, Richard Mayeux, Jennifer J. Manly, Nicole Schupf, and Jose A. Luchsinger, "Mediterranean Diet and Mild Cognitive Impairment," *Archives of Neurology* 66, no. 2 (2009): 216–25.

3. Nikolaos Scarmeas, Jose A. Luchsinger, Nicole Schupf, Adam M. Brickman, Stephanie Cosentino, Ming X. Tang, and Yaakov Stern, "Physical Activity, Diet, and Risk of Alzheimer Disease," *Journal of the American Medical Association* 302, no. 6 (2009): 627–37.

4. G. L. Bowman, L. C. Silbert, D. Howieson, H. H. Dodge, M. G. Traber, B. Frei, J. A. Kaye, J. Shannon, and J. F. Quinn, "Nutrient Biomarker Patterns, Cognitive Function, and MRI Measures of Brain Aging," *Neurology* 78, no. 4 (2012): 241–49.

5. R. Agrawal and F. Gómez-Pinilla, "'Metabolic Syndrome' in the Brain: Deficiency in Omega-3 Fatty Acid Exacerbates Dysfunctions in Insulin Receptor Signalling and Cognition," *Journal of Physiology* 590, part 10 (2012): 2485–99.

6. R. Molteni, R. J. Barnard, Z. Ying, C. K. Roberts, and F. Gómez-Pinilla, "A High-Fat, Refined-Sugar Diet Reduces Hippocampal Brain-Derived Neurotrophic Factor, Neuronal Plasticity, and Learning," *Neuroscience* 112, no. 4 (2002): 803–14. S. Sharma, Y. Zhuang, and F. Gómez-Pinilla, "High-Fat Diet Transition Reduces Brain DHA Levels Associated with Altered Brain Plasticity and Behaviour," *Scientific Reports* 2 (2012): 431.

7. www.brain-armor.com/benefits/cardiovascular.aspx.

8. Annalien Dalton, Petronella Wolmarans, Regina Witthuhn, Martha van Stuijvenberg, Sonja Swanevelder, and Cornelius Smuts, "A Randomized Control Trial in Schoolchildren Showed Improvement in Cognitive Function After Consuming a Bread Spread, Containing Fish Flour from a Marine Source," *Prostaglandins, Leukotrienes, and Essential Fatty Acids* 80, nos. 2–3 (2009): 143–49.

9. J. L. Kim, A. Winkvist, M. A. Åberg, N. Åberg, R. Sundberg, K. Torén, and Jo Brisman, "Fish Consumption and School Grades in Swedish Adolescents: A Study of the Large General Population," *Acta Paediatrica* 99, no. 1 (2010): 72–77.

10. M. F. Muldoon, C. M. Ryan, L. Sheu, J. K. Yao, S. M. Conklin, and S. B. Manuck, "Serum Phospholipid Docosahexaenoic Acid Is Associated with Cognitive Functioning During Middle Adulthood," *Journal of Nutrition* 140, no. 4 (2010): 848–53.

11. K. Yurko-Mauro, D. McCarthy, D. Rom, E. B. Nelson, A. S. Ryan, A. Blackwell, N. Salem Jr., M. Stedman, and MIDAS Investigators, "Beneficial Effects of Docosahexaenoic Acid on Cognition in Age-Related Cognitive Decline," *Alzheimer's and Dementia* 6, no. 6 (2010): 456–64.

12. Majid Fotuhi, Payam Mohassel, and Kristine Yaffe, "Fish Consumption, Long-chain Omega-3 Fatty Acids and Risk of Cognitive Decline or Alzheimer Disease: A Complex Association," *Nature Clinical Practice Neurology* 5, no. 3 (2009): 140–52.

13. Veronica Witte, Lucia Kerti, and Agnes Flöel, "Effects of Omega-3 Supplementation on Brain Structure and Function in Healthy Elderly Subjects," *Alzheimer's and Dementia* 8, no. 4 suppl. (2012): P69.

14. J. Mark Davis, E. Angela Murphy, Martin D. Carmichael, and Ben Davis, "Quercetin Increases Brain and Muscle Mitochondrial Biogenesis and Exercise Tolerance," *American Journal of Physiology* 296, no. 4 (2009): R1071–77.

15. J. Mark Davis, Catherine J. Carlstedt, Stephen Chen, Martin D. Carmichael, and E. Angela Murphy, "The Dietary Flavonoid Quercetin Increases VO_2 Max and Endurance Capacity," *International Journal of Sport Nutrition and Exercise Metabolism* 20, no. 1 (2010): 56–62.

16. Catarina Rendeiro, David Vauzour, Rebecca J. Kean, Laurie T. Butler, Marcus Rattray, Jeremy P. E. Spencer, and Claire M. Williams, "Blueberry Supplementation Induces Spatial Memory Improvements and Region-Specific Regulation of Hippocampal BDNF mRNA Expression in Young Rats," *Psychopharmacology* 223, no. 3 (2012): 319–30.

17. Gregory A. Moy and Ewan C. McNay, "Caffeine Prevents Weight Gain and Cognitive Impairment Caused by a High-Fat Diet While Elevating Hippocampal BDNF," *Physiology and Behavior* 109 (2013): 69–74.

18. Karin Ried, Thomas R. Sullivan, Peter Fakler, Oliver R. Frank, and Nigel P. Stocks, "Effect of Cocoa on Blood Pressure," *Cochrane Database of Systematic Reviews* 8, art. no. CD008893 (2012).

19. Franz H. Messerli, "Chocolate Consumption, Cognitive Function, and Nobel Laureates," *New England Journal of Medicine* 367, no. 16 (2012): 1562–64.

20. Giselle P. Lim, Teresa Chu, Fusheng Yang, Walter Beech, Sally A. Frautschy, and Greg M. Cole, "The Curry Spice Curcumin Reduces Oxidative Damage and Amyloid Pathology in an Alzheimer Transgenic Mouse," *Journal of Neuroscience* 21, no. 21 (2001): 8370–77.

21. T. Ahmed, S. A. Enam, and A. H. Gilani, "Curcuminoids Enhance Memory in an Amyloid-Infused Rat Model of Alzheimer's Disease," *Neuroscience* 169, no. 3 (2010): 1296–306.

22. James A. Joseph, Barbara Shukitt-Hale, and Lauren M. Willis, "Grape Juice, Berries, and Walnuts Affect Brain Aging and Behavior," *Journal of Nutrition* 139, no. 9 (2009): 1813S–17S.

23. M. Rahvar, M. Nikseresht, S. M. Shafiee, F. Naghibalhossaini, M. Rasti, M. R. Panjehshahin, and A. A. Owji, "Effect of Oral Resveratrol on the BDNF Gene Expression in the Hippocampus of the Rat Brain," *Neurochemical Research* 36, no. 5 (2011): 761–65.

24. M. Claire Cartford, Carmelina Gemma, and Paula C. Bickford, "Eighteen-Month-Old Fischer 344 Rats Fed a Spinach-Enriched Diet Show Improved Delay Classical Eyeblink Conditioning and Reduced Expression of Tumor Necrosis Factor Alpha (TNFalpha) and TNFbeta in the Cerebellum," *Journal of Neuroscience* 22, no. 14 (2002): 5813–16.

25. A. C. Nobre, A. Rao, and G. N. Owen, "L-Theanine, a Natural Constituent in Tea, and Its Effect on Mental State," *Asia Pacific Journal of Clinical Nutrition* 17, supplement 1 (2008): 167–68.

26. C. Wakabayashi, T. Numakawa, M. Ninomiya, S. Chiba, and H. Kunugi, "Behavioral and Molecular Evidence for Psychotropic Effects in L-Theanine," *Psychopharmacology* 219, no. 4 (2012): 1099–109.

27. A. Takeda, K. Sakamoto, H. Tamano, K. Fukura, N. Inui, S. W. Suh, S. J. Won, and H. Yokogoshi, "Facilitated Neurogenesis in the Developing Hippocampus After Intake of Theanine, an Amino Acid in Tea Leaves, and Object Recognition Memory," *Cellular and Molecular Neurobiology* 31, no. 7 (2011): 1079–88.

28. Y. Wang, M. Li, X. Xu, M. Song, H. Tao, and Y. Bai, "Green Tea Epigallocatechin-3-Gallate (EGCG) Promotes Neural Progenitor Cell Proliferation and Sonic Hedgehog Pathway Activation During Adult Hippocampal Neurogenesis," *Molecular Nutrition and Food Research* 56, no. 8 (2012): 1292–303.

29. Jiong Yan, Zhihui Feng, Jia Liu, Weili Shen, Ying Wang, Karin Wertz, Peter Weber, Jiangang Long, and Jiankang Liu, "Enhanced Autophagy Plays a Cardinal Role in Mitochondrial Dysfunction in Type 2 Diabetic Goto-Kakizaki (GK) Rats: Ameliorating Effects of (-)-Epigallocatechin-3-Gallate," *Journal of Nutritional Biochemistry* 23, no. 7 (2012): 716–24.

30. Jouni Karppi, Jari Laukkanen, Juhani Sivenius, Kimmo Ronkainen, and Sudhir Kurl, "Serum Lycopene Decreases the Risk of Stroke in Men," *Neurology* 79, no. 15 (2012): 1540–47.

31. Roger Ho, Mike Cheung, Erin Fu, Hlaing Win, Min Zaw, Amanda Ng, and Anselm Mak, "Is High Homocysteine Level a Risk Factor for Cognitive Decline in Elderly? A Systematic Review, Meta-Analysis, and Meta-Regression," *American Journal of Geriatric Psychiatry* 19, no. 7 (2011): 607–17.

32. T. den Heijer, S. E. Vermeer, R. Clarke, M. Oudkerk, P. J. Koudstaal, A. Hofman, and M. M. B. Breteler, "Homocysteine and Brain Atrophy on MRI of Non-Demented Elderly," *Brain* 126 (2003): 170–75.

33. Michael J. Firbank, Sunil K. Narayan, Brian K. Saxby, Gary A. Ford, and John T. O'Brien, "Homocysteine Is Associated with Hippocampal and White Matter Atrophy in Older Subjects with Mild Hypertension," *International Psychogeriatrics* 22, no. 5 (2010): 804–11.

Chapter 6: The Path to a Calmer, Sharper Brain

1. Ye-Ha Jung, Do-Hyung Kang, Min Soo Byun, Geumsook Shim, Soo Jin Kwon, Go-Eun Jang, Ul Soon Lee, Seung Chan An, Joon Hwan Jang, and Jun Soo Kwon, "Influence of Brain-Derived Neurotrophic Factor and Catechol O-Methyl Transferase Polymorphisms on Effects of Meditation on Plasma Catecholamines and Stress," *Stress* 15, no. 1 (2012): 97–104.

2. G. L. Xiong and P. M. Doraiswamy, "Does Meditation Enhance Cognition and Brain Plasticity?" *Annals of the New York Academy of Sciences* 1172 (2009): 63–69.

3. A. B. Newberg, N. Wintering, M. R. Waldman, D. Amen, D. S. Khalsa, and A. Alavi, "Cerebral Blood Flow Differences Between Long-Term Meditators and Non-Meditators," *Conscious and Cognition* 19, no. 4 (2010): 899–905.

4. J. Brefczynski-Lewis, A. Lutz, H. S. Schaefer, D. B. Levinson, and R. Davidson, "Neural Correlates of Attentional Expertise in Long-Term Meditation Practitioners," *Proceedings of the National Academy of Sciences* 104, no. 27 (2007): 11483–88.

5. Heleen A. Slagter, Antoine Lutz, Lawrence L. Greischar, Andrew D. Francis, Sander Nieuwenhuis, James M. Davis, and Richard J. Davidson, "Mental Training Affects Distribution of Limited Brain Resources," *PLoS Biology* 5, no. 6 (2007): E138.

6. Sara W. Lazar, Catherine E. Kerr, Rachel H. Wasserman, Jeremy R. Gray, Douglas N. Greve, Michael T. Treadway, Metta McGarvey, Brian T. Quinn, Jeffery A. Dusek,

Herbert Benson, Scott L. Rauch, Christopher I. Moore, and Bruce Fischl, "Meditation Experience Is Associated with Increased Cortical Thickness," *NeuroReport* 16, no. 17 (2005): 1893–97.

7. Britta K. Hölzel, James Carmody, Mark Vangel, Christina Congleton, Sita M. Yerramsetti, Tim Gard, and Sara W. Lazar, "Mindfulness Practice Leads to Increases in Regional Brain Gray Matter Density," *Psychiatry Research: Neuroimaging* 191, no. 1 (2011): 36–43.

8. Eileen Luders, Arthur W. Toga, Natasha Lepore, and Christian Gaser, "The Underlying Anatomical Correlates of Long-Term Meditation: Larger Hippocampal and Frontal Volumes of Gray Matter," *NeuroImage* 45, no. 3 (2009): 672–78.

9. Eileen Luders, Kristi Clark, Katherine L. Narr, and Arthur W. Toga, "Enhanced Brain Connectivity in Long-Term Meditation Practitioners," *NeuroImage* 57, no. 4 (2011): 1308–16.

10. Jim Robbins, *A Symphony in the Brain* (New York: Grove Press, 2000).

11. David Vernon, Tobias Egner, Nick Cooper, Theresa Compton, Claire Neilands, Amna Sheri, and John Gruzelier, "The Effect of Training Distinct Neurofeedback Protocols on Aspects of Cognitive Performance," *International Journal of Psychophysiology* 47, no. 1 (2003): 75–85.

12. Benedikt Zoefel, René J. Huster, and Christoph S. Herrmann, "Neurofeedback Training of the Upper Alpha Frequency Band in EEG Improves Cognitive Performance," *NeuroImage* 54, no. 2 (2011): 1427–31.

13. Tomas Ros, Merrick J. Moseley, Philip A. Bloom, Larry Benjamin, Lesley A. Parkinson, and John H. Gruzelier, "Optimizing Microsurgical Skills with EEG Neurofeedback," *BMC Neuroscience* 10, no. 87 (2009): 83.

14. C. Escolano, M. Aguilar, and J. Minguez, "EEG-Based Upper Alpha Neurofeedback Training Improves Working Memory Performance," proceedings of the Annual International Conference of the IEEE Engineering in Medicine and Biology Society (2011): 2327–30.

15. Tiffany Field, Miguel Diego, and Maria Hernandez-Reif, "Tai Chi / Yoga Effects on Anxiety, Heartrate, EEG, and Math Computations," *Complementary Therapies in Clinical Practice* 16, no. 4 (2010): 235–38.

16. Tsutomu Kamei, Yoshitaka Toriumi, Hiroshi Kimura, Satoshi Ohno, Hiroaki Kumano, and Keishin Kimura, "Decrease in Serum Cortisol During Yoga Exercise Is Correlated with Alpha Wave Activation," *Perceptual and Motor Skills* 90, no. 3, part 1 (2000): 1027–32.

17. Hazem Doufesh, Tarig Faisal, Kheng-Seang Lim, and Fatimah Ibrahim, "EEG Spectral Analysis on Muslim Prayers," *Applied Psychophysiology and Biofeedback* 37, no. 1 (2012): 11–18.

18. Eliezer Schnall, Sylvia Wassertheil-Smoller, Charles Swencionis, Vance Zemon, Lesley Tinker, Mary Jo O'Sullivan, Linda Van Horn, and Mimi Goodwin, "The Relationship Between Religion and Cardiovascular Outcomes and All-Cause Mortality in the Women's Health Initiative Observational Study," *Psychology and Health* 25, no. 2 (2010): 249–63.

Chapter 7: Building Brain "Muscles"

1. Emma G. Duerden and Danièle Laverdure-Dupont, "Practice Makes Cortex," *Journal of Neuroscience* 28, no. 35 (2008): 8655–57.

2. J. Mårtensson, J. Eriksson, N. C. Bodammer, M. Lindgren, M. Johansson, L. Nyberg, and M. Lövdén, "Growth of Language-Related Brain Areas After Foreign Language Learning," *NeuroImage* 63, no. 1 (2012): 240–44.

3. Bogdan Draganski, Christian Gaser, Gerd Kempermann, H. Georg Kuhn, Jürgen Winkler, Christian Büchel, and Arne May, "Temporal and Spatial Dynamics of Brain Structure Changes During Extensive Learning," *Journal of Neuroscience* 26, no. 23 (2006): 6314–17.

4. Eleanor A. Maguire, David G. Gadian, Ingrid S. Johnsrude, Catriona D. Good, John Ashburner, Richard S. J. Frackowiak, and Christopher D. Frith, "Navigation-Related Structural Change in the Hippocampi of Taxi Drivers," *Proceedings of the National Academy of Science of the United States of America* 97, no. 8 (2000): 4398–403.

5. Katherine Woollett and Eleanor A. Maguire, "Acquiring 'The Knowledge' of London's Layout Drives Structural Brain Changes," *Current Biology* 21, no. 24 (2011): 2109–14.

6. Madeleine Fortin, Patrice Voss, Catherine Lord, Maryse Lassonde, Jens Pruessner, Dave Saint-Amour, Constant Rainville, and Franco Lepore, "Wayfinding in the Blind: Larger Hippocampal Volume and Supranormal Spatial Navigation," *Brain* 131, no. 11 (2008): 2995–3005.

7. Bogdan Draganski, Christian Gaser, Volker Busch, Gerhard Schuierer, Ulrich Bogdahn, and Arne May, "Neuroplasticity: Changes in Grey Matter Induced by Training," *Nature* 427, no. 6972 (2004): 311–12.

8. Ladina Bezzola, Susan Mérillat, Christian Gaser, and Lutz Jäncke, "Training-Induced Neural Plasticity in Golf Novices," *Journal of Neuroscience* 31, no. 35 (2011): 12444–48.

9. J. Verghese, R. B. Lipton, M. J. Katz. C. B. Hall, C. A. Derby, G. Kuslansky, A. F. Ambrose, M. Sliwinski, and H. Buschke, "Leisure Activities and the Risk of Dementia in the Elderly," *New England Journal of Medicine* 348, no. 25 (2003): 2508–16. R. S. Wilson, L. L. Barnes, N. T. Aggarwal, P. A. Boyle, L. E. Hebert, C. F. Mendes de Leon, and D. A. Evans, "Cognitive Activity and the Cognitive Morbidity of Alzheimer Disease," *Neurology* 75, no. 11 (2010). N. Scarmeas, G. Levy, M. X. Tang, J. Manly, and Y. Stern, "Influence of Leisure Activity on the Incidence of Alzheimer's Disease," *Neurology* 57, no. 12 (2001): 2236–42.

10. Michelle C. Carlson, Jane S. Saczynski, George W. Rebok, Teresa Seeman, Thomas A. Glass, Sylvia McGill, James Tielsch, Kevin D. Frick, Joel Hill, and Linda P. Fried, "Exploring the Effects of an 'Everyday' Activity Program on Executive Function and Memory in Older Adults: Experience Corps," *Gerontologist* 48, no. 6 (2008): 793–801.

11. Michelle C. Carlson, Kirk I. Erickson, Arthur F. Kramer, Michelle W. Voss, Natalie Bolea, Michelle Mielke, Sylvia McGill, George W. Rebok, Teresa Seeman, and Linda P. Fried, "Evidence for Neurocognitive Plasticity in At-Risk Older Adults: The Experience Corps Program," *Journals of Gerontology, Series A: Biological Sciences and Medical Sciences* 64, no. 12 (2009): 1275–82.

12. www.geocaching.com.

Chapter 8: Ready, Set, Go

1. Paul D. Loprinzi and Bradley J. Cardinal, "Association Between Objectively Measured Physical Activity and Sleep NHANES 2005–2006," *Mental Health and Physical Activity* 4, no. 2 (2011): 65–69.

Chapter 9: Shrinking the Brain One Night at a Time

1. www.cdc.gov/features/dssleep.

2. Ying He, Christopher R. Jones, Nobuhiro Fujiki, Ying Xu, Bin Guo, Jimmy L. Holder Jr., Moritz J. Rossner, Seiji Nishino, and Ying-Hui Fu, "The Transcriptional Repressor DEC2 Regulates Sleep Length in Mammals," *Science* 325, no. 5942 (2009): 866–70.

3. Anton Sirota, Jozsef Csicsvari, Derek Buhl, and György Buzsáki, "Communication Between Neocortex and Hippocampus During Sleep in Rodents," *PNAS* 100, no. 4 (2003): 2065–69.

4. Lisa Marshall and Jan Born, "The Contribution of Sleep to Hippocampus-Dependent Memory Consolidation," *Trends in Cognitive Sciences* 11, no. 10 (2007): 442–50.

5. www.sleepapnea.org/i-am-a-health-care-professional.html.

6. Karl A. Franklin, Carin Sahlin, Hans Stenlund, and Eva Lindberg, "Sleep Apnoea Is a Common Occurrence in Females," *European Respiratory Journal* 41, no. 3 (2013): 610–15.

7. F. Chung, R. Subramanyam, P. Liao, E. Sasaki, C. Shapiro, and Y. Sun, "High STOP-Bang Score Indicates a High Probability of Obstructive Sleep Apnea," *British Journal of Anaesthesia* 108, no. 5 (2012): 768–75.

8. Paul M. Macey, Luke A. Henderson, Katherine E. Macey, Jeffry R. Alger, Robert C. Frysinger, Mary A. Woo, Rebecca K. Harper, Frisca L. Yan-Go, and Ronald M. Harper, "Brain Morphology Associated with Obstructive Sleep Apnea," *American Journal of Respiratory and Critical Care Medicine,* 166, no. 10 (2002): 1382–87.

9. http://newsroom.heart.org/news/sleep-apnea-linked-to-silent-strokes-221516.

10. W.-H. Wang, G.-P. He, X.-P. Xiao, C. Gu, and H.-Y. Chen, "Relationship Between Brain-Derived Neurotrophic Factor and Cognitive Function of Obstructive Sleep Apnea/Hypopnea Syndrome Patients," *Asia Pacific Journal of Tropical Medicine* 5, no. 11 (2012): 906–10.

11. Nicola Canessa, Vincenza Castronovo, Stefano F. Cappa, Mark S. Aloia, Sara Marelli, Andrea Falini, Federica Alemanno, and Luigi Ferini-Strambi, "Obstructive Sleep Apnea: Brain Structural Changes and Neurocognitive Function Before and After Treatment," *American Journal of Respiratory and Critical Care Medicine* 183, no. 10 (2011): 1419–26.

12. Paul M. Macey, Rajesh Kumar, Frisca L. Yan-Go, Mary A. Woo, and Ronald M. Harper, "Sex Differences in White Matter Alterations Accompanying Obstructive Sleep Apnea," *Sleep* 35, no 12 (2012): 1603–13.

13. Liat Ayalon, Sonia Ancoli-Israel, and Sean P. A. Drummond, "Obstructive Sleep Apnea and Age: A Double Insult to Brain Function?" *American Journal of Respiratory and Critical Care Medicine* 182, no. 3 (2010): 413–19.

14. Khalid Yaouhi, Françoise Bertran, Patrice Clochon, Florence Mézenge, Pierre Denise, Jean Foret, Francis Eustanche, and Béatrice Desgranges, "A Combined

Neuropsychological and Brain Imaging Study of Obstructive Sleep Apnea," *Journal of Sleep Research* 18, no. 1 (2009): 36–48.

15. Randy F. Crossland, David J. Durgan, Eric E. Lloyd, Sharon C. Phillips, Sean P. Marrelli, and Robert M. Bryan, "Cerebrovascular Consequences of Obstructive Sleep Apnea," abstract presented at the American Physiological Society's Experimental Biology conference in April 2012.

16. M. L. Perlis, H. Merica, M. T. Smith, and D. E. Giles, "Beta EEG Activity and Insomnia," *Sleep Medicine Reviews* 5, no. 5 (2001): 365–76.

17. Thomas C. Neylan, Susanne G. Mueller, Zhen Wang, Thomas J. Metzler, Maryann Lenoci, Diana Truran, Charles R. Marmar, Michael W. Weiner, and Norbert Schuff, "Insomnia Severity Is Associated with a Decreased Volume of the CA3/Dentate Gyrus Hippocampal Subfield," *Biological Psychiatry* 68, no. 5 (2010): 494–96.

18. C. Mirescu, J. D. Peters, L. Noiman, and E. Gould, "Sleep Deprivation Inhibits Adult Neurogenesis in the Hippocampus by Elevating Glucocorticoids," *Proceedings of the National Academy of Sciences of the United States of America* 103, no. 50 (2006): 19170–75.

19. Presented at Sleep 2012, the annual meeting of the Associated Professional Sleep Societies.

20. Presented at the Alzheimer's Association International Conference 2012.

21. Christian Benedict, Samantha J. Brooks, Owen G. O'Daly, Markus S. Almèn, Arvid Morell, Karin Åberg, Malin Gingnell, Bernd Schultes, Manfred Hallschmid, Jan-Erik Broman, Elna-Marie Larsson, and Helgi B. Schiöth, "Acute Sleep Deprivation Enhances the Brain's Response to Hedonic Food Stimuli: An fMRI Study," *Journal of Clinical Endocrinology and Metabolism* 97, no. 3 (2012): E443–47.

22. Y. Wang, J.-J. Wang, M.-Q. Zhao, S.-M. Liu, and L.-Z. Li, "Changes of Serum Brain-Derived Neurotrophic Factor in Children with Obstructive Sleep Apnoea-Hypopnoea Syndrome Following Adenotonsillectomy," *Journal of International Medical Research* 38, no. 6 (2010): 1942–51.

23. Melvi Methippara, Tariq Bashir, Natalia Suntsova, Ron Szymusiak, and Dennis McGinty, "Hippocampal Adult Neurogenesis Is Enhanced by Chronic Eszopiclone Treatment in Rats," *Journal of Sleep Research* 19, no. 3 (2010): 384–93.

Chapter 10: How Stress or Depression May Be Shrinking Your Brain

1. Roman Duncko, Linda Johnson, Kathleen Merikangas, and Christian Grillon, "Working Memory Performance After Acute Exposure to the Cold Pressor Stress in Healthy Volunteers," *Neurobiology of Learning and Memory* 91, no. 4 (2009): 377–81.

2. C. R. Park, A. M. Campbell, and D. M. Diamond, "Chronic Psychosocial Stress Impairs Learning and Memory and Increases Sensitivity to Yohimbine in Adult Rats," *Biological Psychiatry* 50, no. 12 (2001): 994–1004.

3. Robert M. Sapolsky, "Glucocorticoids, Stress, and Their Adverse Neurological Effects: Relevance to Aging," *Experimental Gerontology* 34, no. 6 (1999): 721–32.

4. Peter J. Gianaros, J. Richard Jennings, Lei K. Sheu, Phil J. Greer, Lewis H. Kuller, and Karen A. Matthews, "Prospective Reports of Chronic Life Stress Predict Decreased Grey Matter Volume in the Hippocampus," *NeuroImage* 35, no. 2 (2007): 795–803.

5. V. Mondelli, A. Cattaneo, M. Belvederi Murri, M. Di Forti, R. Handley, N. Hepgul, A. Miorelli, S. Navari, A. S. Papadopoulos, K. J. Aitchison, C. Morgan, R. M. Murray, P. Dazzan, and C. M. Pariante, "Stress and Inflammation Reduce Brain-Derived

Neurotrophic Factor Expression in First-Episode Psychosis: A Pathway to Smaller Hippocampal Volume," *Journal of Clinical Psychiatry* 72, no. 12 (2011): 1677–84.

6. C. Liston, B. McEwen, and B. J. Casey, "Psychosocial Stress Reversibly Disrupts Prefrontal Processing and Attentional Control," *Proceedings of the National Academy of Science of the United States of America* 10, no. 3 (2009): 912–17.

7. Gary Evans and Dana Johnson, "Stress and Open-Office Noise," *Journal of Applied Psychology* 85, no. 5 (2000): 779–83.

8. Gary Evans, Peter Lercher, Markus Meis, Hartmut Ising, and Walter Kofter, "Community Noise Exposure and Stress in Children," *Journal of the Acoustical Society of America* 109, no. 3 (2001): 1023–27.

9. Staffan Hygge, Gary W. Evans, and Monika Bullinger, "A Prospective Study of Some Effects of Aircraft Noise on Cognitive Performance in Schoolchildren," *Psychological Science* 13, no. 5 (2002): 469–74.

10. Poul Videbech and Barbara Ravnkilde, "Hippocampal Volume and Depression: A Meta-Analysis of MRI Studies," *American Journal of Psychiatry* 161, no. 11 (2004): 1957–66.

11. Michael Chen, J. Paul Hamilton, and Ian H. Gotlib, "Decreased Hippocampal Volume in Healthy Girls at Risk of Depression," *Archives of General Psychiatry* 67, no. 3 (2010): 270–76.

12. A. R. Brunoni, M. Lopes, and F. Fregni, "A Systematic Review and Meta-Analysis of Clinical Studies on Major Depression and BDNF Levels: Implications for the Role of Neuroplasticity in Depression," *International Journal of Neuropsychopharmacology* 11, no. 8 (2008): 1169–80.

13. C. J. Bench, K. J. Friston, R. G. Brown, L. C. Scott, R. S. J. Frackowiak, and R. J. Dolan, "The Anatomy of Melancholia: Focal Abnormalities of Cerebral Blood Flow in Major Depression," *Psychological Medicine* 22, no. 3 (1992): 607–15.

14. P. Videbech, "PET Measurements of Brain Glucose Metabolism and Blood Flow in Major Depressive Disorder: A Critical Review," *Acta Psychiatrica Scandanavica* 101, no. 1 (2000): 11–20.

15. Mark Gilbertson, Martha Shenton, Aleksandra Ciszewski, Kiyoto Kasai, Natasha Lasko, Scott Orr, and Roger Pitman, "Smaller Hippocampal Volume Predicts Pathologic Vulnerability to Psychological Trauma," *Nature Neuroscience* 5, no. 11 (2002): 1242–47.

16. M. Boldrini, R. Hen, M. D. Underwood, G. B. Rosoklija, A. J. Dwork, J. J. Mann, and V. Arango, "Hippocampal Angiogenesis and Progenitor Cell Proliferation Are Increased with Antidepressant Use in Major Depression," *Biological Psychiatry* 72, no. 7 (2012): 562–71.

17. R. S. Wilson, L. L. Barnes, C. F. Mendes de Leon, N. T. Aggarwal, J. S. Schneider, J. Bach, J. Pilat, L. A. Beckett, S. E. Arnold, D. A. Evans, and D. A. Bennett, "Depressive Symptoms, Cognitive Decline, and Risk of AD in Older Persons," *Neurology* 59, no. 3 (2002): 364–70.

18. Ross Andel, Michael Crowe, Elizabeth Hahn, James Mortimer, Nancy Pedersen, Laura Fratiglioni, Boo Johansson, and Margaret Gatz, "Work-Related Stress May Increase the Risk of Vascular Dementia," *Journal of the American Geriatric Society* 60, no. 1 (2011): 60–67.

19. Adam Clark, Alexander Seidler, and Michael Miller, "Inverse Association Between Sense of Humor and Coronary Heart Disease," *International Journal of Cardiology* 80, no. 1 (2001): 87–88.

20. M. Miller, C. Mangano, Y. Park, R. Goel, G. D. Plotnick, and R. A. Vogel, "Impact of Cinematic Viewing on Endothelial Function," *Heart* 92, no. 2 (2006): 261–62.

21. Mary Payne Bennett and Cecile Lengacher, "Humor and Laughter May Influence Health IV: Humor and Immune Function," *Evidence Based Complementary and Alternative Medicine* 6, no. 2 (2009): 159–64.

Chapter 11: Bigger Belly, Smaller Brain

1. R. Corripio, J. M. Gónzalez-Clemente, J. Pérez-Sanchez, S. Näf, L. Gallart, J. Vendrell, and A. Caixàs, "Plasma Brain-Derived Neurotrophic Factor in Prepubertal Obese Children: Results from a Two-Year Lifestyle Intervention Programme," *Clinical Endocrinology* 77, no. 5 (2012): 715–20.

2. K. S. Krabbe, A. R. Nielsen, R. Krogh-Madsen, P. Plomgaard, P. Rasmussen, C. Erikstrup, C. P. Fischer, B. Lindegaard, A. M. W. Petersen, S. Taudorf, N. H. Secher, H. Pilegaard, H. Bruunsgaard, and B. K. Pedersen, "Brain-Derived Neurotrophic Factor (BDNF) and Type 2 Diabetes," *Diabetologia* 50, no. 2 (2007): 431–38.

3. Alex C. Birdsill, Cynthia M. Carlsson, Auriel A. Willette, Ozioma C. Okonkwo, Sterling C. Johnson, Guofan Xu, Jennifer M. Oh, Catherine L. Gallagher, Rebecca L. Koscik, Erin M. Jonaitis, Bruce P. Hermann, Asenath LaRue, Howard A. Rowley, Sanjay Asthana, Mark A. Sager, and Barbara B. Bendlin, "Low Cerebral Blood Flow Is Associated with Lower Memory Function in Metabolic Syndrome," *Obesity*, published online May 19, 2013, doi: 10.1002/oby.20170.

4. Cyrus A. Raji, April J. Ho, Neelroop N. Parikshak, James T. Becker, Oscar L. Lopez, Lewis H. Kuller, Xue Hua, Alex D. Leow, Arthur W. Toga, and Paul M. Thompson, "Brain Structure and Obesity," *Human Brain Mapping* 31, no. 3 (2010): 353–64.

5. Michael A. Ward, Cynthia M. Carlsson, Mehul A. Trivedi, Mark A. Sager, and Sterling C. Johnson, "The Effect of Body Mass Index on Global Brain Volume in Middle-Aged Adults: A Cross-Sectional Study," *BMC Neurology* 5, no. 1 (2004): 23.

6. Victor W. Swayze II, Arnold E. Andersen, Nancy C. Andreasen, Stephan Arndt, Yutaka Sato, and Steve Ziebell, "Brain Tissue Volume Segmentation in Patients with Anorexia Nervosa Before and After Weight Normalization," *International Journal of Eating Disorders* 33, no. 1 (2003): 33–44.

7. Stéphanie Debette, Alexa Beiser, Udo Hoffmann, Charles DeCarli, Christopher J. O'Donnell, Joseph M. Massaro, Rhoda Au, Jayandra J. Himali, Philip A. Wolf, Caroline S. Fox, and Sudha Seshadri, "Visceral Fat Is Associated with Lower Brain Volume in Healthy Middle-Aged Adults," *Annals of Neurology* 68, no. 2 (2010): 136–44.

8. R. A. Whitmer, D. R. Gustafson, E. Barrett-Connor, M. N. Haan, E. P. Gunderson, and K. Yaffe, "Central Obesity and Increased Risk of Dementia More than Three Decades Later," *Neurology* 71, no. 14 (2008): 1057–64.

9. S. Debette, S. Seshadri, A. Beiser, R. Au, J. J. Himali, C. Palumbo, P. A. Wolf, and C. DeCarli, "Midlife Vascular Risk Factor Exposure Accelerates Structural Brain Aging and Cognitive Decline," *Neurology* 77, no. 5 (2011): 461–68. E. S. C. Korf, L. R. White, P. Scheltens, and L. J. Launer, "Brain Aging in Very Old Men with Type 2 Diabetes: The Honolulu-Asia Aging Study," *Diabetes Care* 29, no. 10 (2006): 2268–74.

10. Thomas Jeerakathil, Jeffrey A. Johnson, Scot H. Simpson, and Sumit R. Majumdar, "Short-Term Risk for Stroke Is Doubled in Persons with Newly Treated Type 2 Diabetes Compared with Persons Without Diabetes: A Population-Based Cohort Study," *Stroke* 38, no. 6 (2007): 1739–43.

11. P. L. Yau, D. C. Javier, C. M. Ryan, W. H. Tsui, B. A. Ardenkani, S. Ten, and A. Convit, "Preliminary Evidence for Brain Complications in Obese Adolescents with Type 2 Diabetes Mellitus," *Diabetologia* 53, no. 11 (2010): 2298–306.

12. Hannah Bruehl, Victoria Sweat, Aziz Tirsi, Bina Shah, and Antonio Convit, "Obese Adolescents with Type 2 Diabetes Mellitus Have Hippocampal and Frontal Lobe Volume Reductions," *Neuroscience and Medicine* 2, no. 1 (2011): 34–42.

13. Kristine Yaffe, Elizabeth Barrett-Connor, Feng Lin, and Deborah Grady, "Serum Lipoprotein Levels, Statin Use, and Cognitive Function in Older Women," *Archives of Neurology* 59, no. 3 (2002): 378–84.

14. Miia Kivipelto, Eeva-Liisa Helkala, Mikko P. Laakso, Tuomo Hänninen, Merja Hallikainen, Kari Alhainen, Hilkka Soininen, Jaakko Tuomilehto, and Aulikki Nissinen, "Midlife Vascular Risk Factors and Alzheimer's Disease in Later Life: Longitudinal Population-Based Study," *British Medical Journal* 322, no. 7300 (2001): 1447–51.

15. Archana Singh-Manoux, David Gimeno, Mika Kivimaki, Eric Brunner, and Michael G. Marmot, "Low HDL Cholesterol Is a Risk Factor for Deficit and Decline in Memory in Midlife: The Whitehall II Study," *Arteriosclerosis, Thrombosis, and Vascular Biology* 28, no. 8 (2008): 1556–62.

16. Marie-Laure Ancelin, Isabelle Carrière, Pascale Barberger-Gateau, Sophie Auriacombe, Olivier Rouaud, Spiros Fourlanos, Claudine Berr, Anne-Marie Dupuy, and Karen Ritchie, "Lipid Lowering Agents, Cognitive Decline, and Dementia: The Three-City Study," *Journal of Alzheimer's Disease* 30, no. 3 (2012): 629–37.

17. Carlos H. Rojas-Fernandez and Jean-Christy F. Cameron, "Is Statin-Associated Cognitive Impairment Clinically Relevant? A Narrative Review and Clinical Recommendations," *Annals of Pharmacotherapy* 46, no. 4 (2012): 549–57.

18. Lauri Nummenmaa, Jussi Hirvonen, Jarna C. Hannukainen, Heidi Immonen, Markus M. Lindroos, Paulina Salminen, and Pirjo Nuutila, "Dorsal Striatum and Its Limbic Connectivity Mediate Abnormal Anticipatory Reward Processing in Obesity," *PLoS ONE* 7, no. 2 (2012): e31089.

19. John Gunstad, Gladys Strain, Michael J. Devlin, Rena Wing, Ronald A. Cohen, Robert H. Paul, Ross D. Crosby, and James E. Mitchell, "Improved Memory Function Twelve Weeks After Bariatric Surgery," *Surgery for Obesity and Related Diseases* 7, no. 4 (2011): 465–72.

Chapter 12: Brain Attacks, Large and Small

1. R. M. Wiseman, B. K. Saxby, E. J. Burton, R. Barber, G. A. Ford, and J. T. O'Brien, "Hippocampal Atrophy, Whole Brain Volume, and White Matter Lesions in Older Hypertensive Subjects," *Neurology* 63, no. 10 (2004): 1892–97.

2. Miia Kivipelto, Eeva-Liisa Helkala, Mikko P. Laakso, Tuomo Hänninen, Merja Hallikainen, Kari Alhainen, Hilkka Soininen, Jaakko Tuomilehto, and Aulikki Nissinen, "Midlife Vascular Risk Factors and Alzheimer's Disease in Later Life: Longitudinal Population-Based Study," *British Medical Journal* 322, no. 7300 (2001): 1447–51.

3. S. Debette, S. Seshadri, A. Beiser, R. Au, J. J. Himali, C. Palumbo, P. A. Wolf, and C. DeCarli, "Midlife Vascular Risk Factor Exposure Accelerates Structural Brain Aging and Cognitive Decline," *Neurology* 77, no. 5 (2011): 461–68.

4. Didier Leys, Hilde Hénon, Marie-Anne Mackowiak-Cordoliani, and Florence Pasquier, "Post-Stroke Dementia," *The Lancet Neurology* 4, no. 11 (2005): 752–59.

5. Juan C. Troncoso, Alan B. Zonderman, Susan M. Resnick, Barbara Crain, Olga Pletnikova, and Richard J. O'Brien, "Effect of Infarcts on Dementia in the Baltimore Longitudinal Study of Aging," *Annals of Neurology* 64, no. 2 (2008): 168–76.

6. National Stroke Association, www.stroke.org.

7. C. DeCarli, D. G. Murphy, M. Tranh, C. L. Grady, J. V. Haxby, J. A. Gillette, J. A. Salerno, A. Gonzales-Aviles, B. Horwitz, and S. I. Rapoport, "The Effect of White Matter Hyperintensity Volume on Brain Structure, Cognitive Performance, and Cerebral Metabolism of Glucose in Fifty-One Healthy Adults," *Neurology* 45, no. 11 (1995): 2077–84.

8. Brett M. Kissela, Jane C. Khoury, Kathleen Alwell, Charles J. Moomaw, Daniel Woo, Opeolu Adeoye, Matthew L. Flaherty, Pooja Khatri, Simona Ferioli, Felipe De Los Rios La Rosa, Joseph P. Broderick, and Dawn O. Kleindorfer, "Age at Stroke: Temporal Trends in Stroke Incidence in a Large, Biracial Population," *Neurology* 79, no. 17 (2012): 1781–87.

9. www.stroke.org/site/DocServer/Scorecard.Q._08.pdf?docID=601.

10. J. A. Egido, O. Castillo, B. Roig, I. Sanz, M. R. Herrero, M. T. Garay, A. M. Garcia, M. Fuentes, and C. Fernandez, "Is Psycho-Physical Stress a Risk Factor for Stroke? A Case-Control Study," *Journal of Neurology, Neurosurgery, and Psychiatry* 83, no. 11 (2012): 1104–10, doi: 10.1136/jnnp-2012-302420.

11. A. M. Bernstein, L. de Koning, A. J. Flint, K. M. Rexrode, and W. C. Willett, "Soda Consumption and the Risk of Stroke in Men and Women," *American Journal of Clinical Nutrition* 95, no. 5 (2012): 1190–99.

12. American Thoracic Society (May 20, 2008), "Obstructive Sleep Apnea Causes Earlier Death in Stroke Patients, Study Finds," *ScienceDaily*, www.sciencedaily.com-/releases/2008/05/080518182655.htm.

13. Susan Marzolini, Paul Oh, William McIlroy, and Dina Brooks, "The Effects of an Aerobic and Resistance Exercise Program on Cognition Following Stroke," *Neurorehabilitation and Neural Repair* 27, no. 5 (2013): 392–402.

14. Susanna Larsson, Jarmo Virtamo, and Alicja Wolk, "Chocolate Consumption and Risk of Stroke: A Prospective Cohort of Men and Meta-Analysis," *Neurology* 79, no. 12 (2012): 1223–29.

Chapter 13: The Battered Brain

1. R. L. Blaylock and J. Maroon, "Immunoexcitotoxicity as a Central Mechanism in Chronic Traumatic Encephalopathy: A Unifying Hypothesis," *Surgical Neurology International* 2 (2011): 107.

2. D. K. Ragan, R. McKinstry, T. Benzinger, J. R. Leonard, and J. A. Pineda, "Alterations in Cerebral Oxygen Metabolism After Traumatic Brain Injury in Children," *Journal of Cerebral Blood Flow Metabolism* 33, no. 1 (2013): 48–52.

3. E. D. Bigler, D. D. Blatter, C. V. Anderson, S. C. Johnson, S. D. Gale, R. O. Hopkins, and B. Burnett, "Hippocampal Volume in Normal Aging and Traumatic Brain Injury," *American Journal of Neuroradiology* 18, no. 1 (1997): 11–23. Mar Ariza, Josep Serra-Grabulosa, Carme Junqué, Blanca Ramirez, Maria Mataro, Antonia Poca, Nuria Bargalló, and Juan Sahuquillo, "Hippocampal Head Atrophy After Traumatic Brain Injury," *Neuropsychologia* 44, no. 1 (2006): 1956–61.

4. Majid Fotuhi, David Do, and Clifford Jack, "Modifiable Factors That Alter the Size of the Hippocampus with Aging," *Nature Reviews Neurology* 8, no. 4 (2012): 189–202.

5. M. H. Beauchamp, M. Ditchfield, J. J. Maller, C. Catroppa, C. Godfrey, J. V. Rosenfeld, M. J. Kean, and V. A. Anderson, "Hippocampus, Amygdala, and Global Brain Changes Ten Years After Childhood Traumatic Brain Injury," *International Journal of Developmental Neuroscience* 29, no. 2 (2011): 137–43.

6. S. P. Broglio, M. B. Pontifex, P. O'Connor, and C. H. Hillman, "The Persistent Effects of Concussion on Neuroelectric Indices of Attention," *Journal of Neurotrauma* 26, no. 9 (2009): 1463–70.

7. Jacob J. Sosnoff, Steven P. Broglio, Sunghoon Shin, and Michael S. Ferrara, "Previous Mild Traumatic Brain Injury and Postural-Control Dynamics," *Journal of Athletic Training* 46, no. 1 (2011): 85–91.

8. Evan L. Breedlove, Meghan Robinson, Thomas M. Talavage, Katherine E. Morigaki, Umit Yoruk, Kyle O'Keefe, Jeff King, Larry J. Leverenz, Jeffrey W. Gilger, and Eric A. Nauman, "Biomechanical Correlates of Symptomatic and Asymptomatic Neurophysiological Impairment in High School Football," *Journal of Biomechanics* 45, no. 7 (2012): 1265–72.

9. Kevin M. Guskiewicz, Michael McCrea, Stephen W. Marshall, Robert C. Cantu, Christopher Randolph, William Barr, James A. Onate, and James P. Kelly, "Cumulative Effects Associated with Recurrent Concussion in Collegiate Football Players: The NCAA Concussion Study," *Journal of the American Medical Association* 290, no. 19 (2003): 2549–55.

10. Kevin M. Guskiewicz, Stephen W. Marshall, Julian Bailes, Michael McCrea, Robert C. Cantu, Christopher Randolph, and Barry D. Jordan, "Association Between Recurrent Concussion and Late-Life Cognitive Impairment in Retired Professional Football Players," *Neurosurgery* 57, no. 4 (2005): 719–26.

11. Steven T. DeKosky, Milos D. Ikonomovic, and Sam Gandy, "Traumatic Brain Injury: Football, Warfare, and Long-Term Effects," *Minnesota Medicine* 93, no. 12 (2010): 46–47.

12. www.cnn.com/2013/01/22/health/cte-study/index.html?hpt=hp_c2.

13. C. M. Baugh, J. M. Stamm, D. O. Riley, B. E. Gavett, M. E. Shenton, A. Lin, C. J. Nowinski, R. C. Cantu, A. C. McKee, and R. A. Stern, "Chronic Traumatic Encephalopathy: Neurodegeneration Following Repetitive Concussive and Subconcussive Brain Trauma," *Brain Imaging Behavior* 6, no. 2 (2012): 244–54.

14. As reported by CNN, the *New York Times*, and other publications.

15. www.aans.org/Patient%20Information/Conditions%20and%20Treatments/Sports-Related%20HeadInjury.aspx.

16. D. W. Wright, A. L. Kellermann, V. S. Hertzberg, P. L. Clark, M. Frankel, F. C. Goldstein, J. P. Salomone, L. L. Dent, O. A. Harris, D. S. Ander, D. W. Lowery, M. M. Patel, D. D. Denson, A. B. Gordon, M. M. Wald, S. Gupta, S. W. Hoffman, and D. G. Stein, "ProTECT: A Randomized Clinical Trial of Progesterone for Acute Traumatic Brain Injury," *Annals of Emergency Medicine* 49, no. 4 (2007): 391–402.

17. J. D. Mills, J. E. Bailes, C. L. Sedney, H. Hutchins, and B. Sears, "Omega-3 Fatty Acid Supplementation and Reduction of Traumatic Axonal Injury in a Rodent Head Injury Model," *Journal of Neurosurgery* 114, no. 1 (2011): 77–84.

Chapter 14: Washing Away Your Brain's Neighborhoods and Highways

1. A. Ruitenberg, J. van Swieten, J. Witteman, K. Mehta, C. van Duijn, A. Hofman, and M. Breteler, "Alcohol Consumption and Risk of Dementia: The Rotterdam Study," *The Lancet* 359, no. 9303 (2002): 281–86.

2. S. Gupta and J. Warner, "Alcohol-Related Dementia: A Twenty-First-Century Silent Epidemic?" *British Journal of Psychiatry* 193, no. 5 (2008): 351–53.

3. R. Momenan, L. E. Steckler, Z. S. Saad, S. van Rafelghem, M. J. Kerich, and D. W. Hommer, "Effects of Alcohol Dependence on Cortical Thickness as Determined by Magnetic Resonance Imaging," *Psychiatry Research* 204, nos. 2–3 (2012): 101–11.

4. A. P. Le Berre, G. Rauchs, R. La Joie, F. Mézenge, C. Boudehent, F. Vabret, S. Segobin, F. Viader, P. Allain, F. Eustache, A. L. Pitel, and H. Beaunieux, "Impaired Decision Making and Brain Shrinkage in Alcoholism," *European Psychiatry* S00924-9338, no. 12 (2012): 136–38.

5. Kristi Reynolds, Brian Lewis, John David L. Nolen, Gregory L. Kinney, Bhavani Sathya, and Jiang He, "Alcohol Consumption and Risk of Stroke," *Journal of the American Medical Association* 289, no. 5 (2003): 579–88. George Winokur, "Alcoholism and Depression," *Substance and Alcohol Actions/Misuse* 4, nos. 2–3 (1983): 111–19.

6. http://pubs.niaaa.nih.gov/publications/aa68/aa68.htm.

7. María Parada, Montserrat Corral, Nayara Mota, Alberto Crego, Socorro Rodríguez Holguín, and Fernando Cadaveira, "Executive Functioning and Alcohol Binge Drinking in University Students," *Addictive Behaviors* 37, no. 2 (2012): 167–72.

8. Tim McQueeny et al., "Binge Drinking," presented in 2011 to the Research Society on Alcoholism.

9. Simone Kühn, Florian Schubert, and Jürgen Gallinat, "Reduced Thickness of Medial Orbitofrontal Cortex in Smokers," *Biological Psychiatry* 68, no. 11 (2010): 1061–106.

10. T. Demirakca, G. Ende, N. Kämmerer, H. Welzel-Marquez, D. Hermann, A. Heinz, and K. Mann, "Effects of Alcoholism and Continued Abstinence on Brain Volumes in Both Genders," *Alcoholism, Clinical and Experimental Research* 35, no. 9 (2011): 1678–85.

11. J. van Eijk, T. Demirakca, U. Frischknecht, D. Hermann, K. Mann, and G. Ende, "Rapid Partial Regeneration of Brain Volume During the First Fourteen Days of Abstinence from Alcohol," *Alcoholism, Clinical and Experimental Research* 37, no. 1 (2013): 67–74.

12. F. T. Crews and K. Nixon, "Mechanisms of Neurodegeneration and Regeneration in Alcoholism," *Alcohol and Alcoholism* 44, no. 2 (2009): 115–27.

Chapter 15: Is It Alzheimer's?

1. Archana Singh-Manoux, Mika Kivimaki, M. Maria Glymour, Alexis Elbaz, Claudine Berr, Klaus P. Ebmeier, Jane E. Ferrie, and Aline Dugravot, "Timing of Onset of Cognitive Decline: Results from Whitehall II Prospective Cohort Study," *British Medical Journal* 344 (2012): d7622.

2. www.alz.org/documents_custom/2012_facts_figures_fact_sheet.pdf.

3. Lon R. White, Brent J. Small, Helen Petrovitch, G. Webster Ross, Kamal Masaki, Robert D. Abbott, John Hardman, Daron Davis, James Nelson, and William Markesbery, "Recent Clinical-Pathologic Research on the Causes of Dementia in Late Life: Update from the Honolulu-Asia Aging Study," *Journal of Geriatric Psychiatry and Neurology* 18, no. 4 (2005): 224–27. Lenore J. Launer, Timothy M. Hughes, and Lon R. White, "Microinfarcts, Brain Atrophy, and Cognitive Function: The Honolulu-Asia Aging Study Autopsy Study," *Annals of Neurology* 70, no. 5 (2011): 774–80. Ingmar Skoog, Lars Nilsson, Bo Palmertz, Lars-Arne Andreasson, and Alvar Svanborg, "A Population-Based Study of Dementia in Eighty-Five-Year-Olds," *New England Journal of Medicine* 328, no. 3 (1993): 153–58.

4. Miia Kivipelto, Eeva-Liisa Helkala, Mikko P. Laakso, Tuomo Hänninen, Merja Hallikainen, Kari Alhainen, Hilkka Soininen, Jaakko Tuomilehto, and Aulikki Nissinen, "Midlife Vascular Risk Factors and Alzheimer's Disease in Later Life: Longitudinal, Population-Based Study," *British Medical Journal* 322, no. 7300 (2001): 1447–51.

5. I. Prohovnik, D. P. Perl, K. L. Davis, L. Libow, G. Lesser, and V. Haroutunian, "Dissociation of Neuropathology from Severity of Dementia in Late-Onset Alzheimer Disease," *Neurology* 66, no. 1 (2006): 49–55.

6. Julie A. Schneider, Zoe Arvanitakis, Woojeong Bang, and David A. Bennett, "Mixed Brain Pathologies Account for Most Dementia Cases in Community-Dwelling Older Persons," *Neurology* 69, no. 24 (2007): 2197–204.

7. K. A. Jellinger and J. Attems, "Neuropathological Evaluation of Mixed Dementia," *Journal of the Neurological Sciences* 257, nos. 1–2 (2007): 80–87.

8. www.healthstudies.umn.edu/nunstudy/faq.jsp.

9. K. M. Langa, N. L. Foster, and E. B. Larson, "Mixed Dementia: Emerging Concepts and Therapeutic Implications," *Journal of the American Medical Association* 292, no. 23 (2004): 2901–8.

10. www.aan.com.

11. H. J. Aizenstein, R. D. Nebes, J. A. Saxton, J. C. Price, C. A. Mathis, N. D. Tsopelas, S. K. Ziolko, J. A. James, B. E. Snitz, P. R. Houck, W. Bi, A. D. Cohen, B. J. Lopresti, S. T. DeKosky, E. M. Halligan, and W. E. Klunk, "Frequent Amyloid Deposition Without Significant Cognitive Impairment Among the Elderly," *Archives of Neurology* 65, no. 11 (2008): 1509–17. K. E. Pike, G. Savage, V. L. Villemagne, S. Ng, S. A. Moss, P. Maruff, C. A. Mathis, W. E. Klunk, C. L. Masters, and C. C. Rowe, "Beta-Amyloid Imaging and Memory in Non-Demented Individuals: Evidence for Preclinical Alzheimer's Disease," *Brain* 130, part 11 (2007): 2837–44. S. N. Gomperts, D. M. Rentz, E. Moran, J. A. Becker, J. J. Locascio, W. E. Klunk, C. A. Mathis, D. R. Elmaleh, T. Shoup, A. J. Fischman, B. T. Hyman, J. H. Growdon, and K. A. Johnson, "Imaging Amyloid Deposition in Lewy Body Diseases," *Neurology* 71, no. 12 (2008): 903–10.

12. Adam S. Fleisher, Kewei Chen, Yakeel T. Quiroz, Laura J. Jakimovich, Madelyn Gutierrez Gomez, Carolyn M. Langois, Jessica B. S. Langbaum, Napatkamon Ayutyanont, Auttawut Roontiva, Pradeep Thiyyagura, Wendy Lee, Hua Mo, Liliana Lopez, Sonia Moreno, Natalia Acosta-Baena, Margarita Giraldo, Gloria Garcia, Rebecca A. Reiman, Matthew J. Kenneth Huentelman, S. Kosik, Pierre N. Tariot, Francisco Lopera, and Eric M. Reiman, "Florbetapir PET Analysis of Amyloid-β Deposition in the Presenilin 1 E280A Autosomal Dominant Alzheimer's Disease Kindred: A Cross-Sectional Study," *The Lancet Neurology* 11, no. 12 (2012): 1057–65.

13. G. W. Small, P. Siddarth, V. Kepe, L. M. Ercoli, A. C. Burggren, S. Y. Bookheimer, K. J. Miller, J. Kim, H. Lavretsky, S.-C. Huang, and J. R. Barrio, "Prediction of Cognitive

Decline by Positron Emission Tomography of Brain Amyloid and Tau," *Archives of Neurology* 69, no. 2 (2012): 215–22.

14. www.cnn.com/2013/01/22/health/cte-study/index.html?hpt=hp_c2.

15. B. Dubois, H. H. Feldman, C. Jacova, S. T. Dekosky, P. Barberger-Gateau, J. Cummings, A. Delacourte, D. Galasko, S. Gauthier, G. Jicha, K. Meguro, J. O'Brien, F. Pasquier, P. Robert, M. Rossor, S. Salloway, Y. Stern, P. J. Visser, and P. Scheltens, "Research Criteria for the Diagnosis of Alzheimer's Disease: Revising the NINCDS-ADRDA Criteria," *The Lancet Neurology* 6, no. 8 (2007): 734–46.

16. D. Erten-Lyons, R. L. Woltjer, H. Dodge, R. Nixon, R. Vorobik, J. F. Calvert, M. Leahy, T. Montine, and J. Kaye, "Factors Associated with Resistance to Dementia Despite High Alzheimer Disease Pathology," *Neurology* 72, no. 4 (2009): 354–60.

17. L. D. Baker, L. L. Frank, K. Foster-Schubert, P. S. Green, C. W. Wilkinson, A. McTiernan, S. R. Plymate, M. A. Fishel, G. S. Watson, B. A. Cholerton, G. E. Duncan, P. D. Mehta, and S. Craft, "Effects of Aerobic Exercise on Mild Cognitive Impairment: A Controlled Trial," *Archives of Neurology* 67, no. 1 (2010): 71–79.

18. L. S. Nagamatsu, T. C. Handy, C. L. Hsu, M. Voss, and T. Liu-Ambrose, "Resistance Training Promotes Cognitive and Functional Brain Plasticity in Seniors with Probable Mild Cognitive Impairment," *Archives of Internal Medicine* 172, no. 8 (2012): 666–68.

19. D. E. Barnes, K. Yaffe, N. Belfor, W. J. Jagust, C. DeCarli, B. R. Reed, and J. H. Kramer, "Computer-Based Cognitive Training for Mild Cognitive Impairment: Results from a Pilot Randomized, Controlled Trial," *Alzheimer Disease and Associated Disorders* 23, no. 3 (2009): 205–10.

20. V. C. Buschert, U. Friese, S. J. Teipel, P. Schneider, W. Merensky, D. Rujescu, H. J. Möller, H. Hampel, and K. Buerger, "Effects of a Newly Developed Cognitive Intervention in Amnestic Mild Cognitive Impairment and Mild Alzheimer's Disease: A Pilot Study," *Journal of Alzheimer's Disease* 25, no. 4 (2011): 679–94.

21. S. S. Simon, J. E. Yokomizo, and C. M. Bottino, "Cognitive Intervention in Amnestic Mild Cognitive Impairment: A Systematic Review," *Neuroscience and Biobehavioral Reviews* 36, no. 4 (2012): 1163–78.

22. www.alz.org/alzheimers_disease_causes_risk_factors.asp.

Chapter 16: Back to the Future

1. N. C. Berchtold and C. W. Cotman, "Evolution in the Conceptualization of Dementia and Alzheimer's Disease: Greco-Roman Period to the 1960s," *Neurobiology of Aging* 19, no. 3 (1998): 173–89.

2. Alois Alzheimer, "Über Eigenartige Krankheitsfälle des Späteren Alters [On Peculiar Cases of Disease at Higher Age]," *Neurologie und Psychiatrie* 4 (1911): 256–86.

3. J. Hardy and G. A. Higgins, "Alzheimer's Disease: The Amyloid Cascade Hypothesis," *Science* 256, no. 5054 (1992): 184–85.

4. I. Prohovnik, D. P. Perl, K. L. Davis, L. Libow, G. Lesser, and V. Haroutunian, "Dissociation of Neuropathology from Severity of Dementia in Late-Onset Alzheimer Disease," *Neurology* 66, no. 1 (2006): 49–55.

5. George M. Savva, Stephen B. Wharton, Paul G. Ince, Gillian Forster, Fiona E. Matthews, and Carol Brayne, "Medical Research Council Cognitive Function and Aging Study: Age, Neuropathology, and Dementia," *New England Journal of Medicine* 360, no. 22 (2009): 2302–9.

6. Richard Robinson, "News from the AAN Annual Meeting: Multiple Pathologies in Many Alzheimer Disease Patients," *Neurology Today* 11, no. 11 (2011): 28–29.

7. Majid Fotuhi, Vladimir Hachinski, and Peter J. Whitehouse, "Changing Perspectives Regarding Late-Life Dementia," *Nature Reviews Neurology* 5, no. 12 (2009): 649–58.

8. Tom Valeo, "Autopsied Brains Show Seniors on Beta Blockers Had Fewer Signs of Dementia," *Neurology Today* 13, no. 5 (2013): 6–8.

9. Denise Head, Julie M. Bugg, Alison M. Goate, Anne M. Fagan, Mark A. Mintun, Tammie Benzinger, David M. Holtzman, and John C. Morris, "Exercise Engagement as a Moderator of the Effects of APOE Genotype on Amyloid Deposition," *Archives of Neurology* 69, no. 5 (2012): 636–43.

10. www.cnn.com/2013/01/22/health/cte-study/index.html?hpt=hp_c2.

Index

Page numbers in *italics* refer to illustrations.